THE CHEMISTRY OF MOOD, MOTIVATION, AND MEMORY

ADVANCES IN BEHAVIORAL BIOLOGY

THE CHEMISTRY OF MOOD, MOTIVATION, AND MEMORY

The proceedings of an interdisciplinary conference on the Chemistry of Mood, Motivation, and Memory held at the University of California, San Francisco, in October 1971

EDITED BY
JAMES L. McGAUGH

Department of Psychobiology
School of Biological Sciences
University of California, Irvine
Irvine, California

ℚ PLENUM PRESS · NEW YORK-LONDON · 1972

Library of Congress Catalog Card Number 72-83801
ISBN 0-306-37904-X

© 1972 Plenum Press, New York
A Division of Plenum Publishing Corporation
227 West 17th Street, New York, N. Y. 10011

United Kingdom edition published by Plenum Press, London
A Division of Plenum Publishing Company, Ltd.
Davis House (4th Floor), 8 Scrubs Lane, Harlesden,
London, NW10 6SE, England

Printed in the United States of America

PREFACE

This volume is based on presentations at an interdisciplinary conference on The Chemistry of Mood, Motivation and Memory which was held at the University of California, San Francisco in October, 1971. The conference was sponsored and supported by the Division of Continuing Education in Health Sciences. We thank Dr. Ruben Dixon and his staff for help in planning the conference and for attending to all of the organizational details.

We particularly thank the participants for their contributions to the conference and for their cooperation in preparing the manuscripts based on their conference presentations.

All of the details involved in preparing the volume for publication were handled by Karen Dodd. We are grateful to her for her tireless, efficient and productive efforts.

We hope that this volume will help to stimulate further interest as well as understanding of the biochemical bases of our behavior.

James L. McGaugh

CONTRIBUTORS

Bernard W. Agranoff, Ph.D.
Department of Psychiatry
Mental Health Research Institute
The University of Michigan
Ann Arbor, Michigan

Samuel H. Barondes, M.D.
Professor of Psychiatry
School of Medicine
University of California at San Diego
La Jolla, California

Barry D. Berger, Ph.D.
Wyeth Laboratories
Philadelphia, Pennsylvania

Enoch Callaway, M.D.
Professor of Psychiatry
University of California
School of Medicine
San Francisco, California

J. Anthony Deutsch, Ph.D.
Professor of Psychology
University of California at San Diego
La Jolla, California

Edward Glassman, Ph.D.
Division of Chemical Neurobiology
Department of Biochemistry
The University of North Carolina
Chapel Hill, North Carolina

Richard Green, M.D.
Department of Psychiatry
University of California
Los Angeles, California

Seymour S. Kety, M.D.
Department of Psychiatry
Harvard Medical School
Director, Psychiatric Research
Harvard Medical School
Boston, Massachusetts

Suzanne Knapp, Ph.D.
Department of Psychiatry
School of Medicine
University of California at San Diego
La Jolla, California

David Krech, Ph.D.
Professor Emeritus
Department of Psychology
University of California
Berkeley, California

Ronald T. Kuczenski, Ph.D.
Department of Psychiatry
School of Medicine
University of California at San Diego
La Jolla, California

Seymour Levine, Ph.D.
Department of Psychiatry
Stanford University
Stanford, California

Morris P. Lipton, M.D.
Department of Psychiatry
School of Medicine
The University of North Carolina
Chapel Hill, North Carolina

Barry Machlus, Ph.D.
Division of Chemical Neurobiology
Department of Biochemistry
The University of North Carolina
Chapel Hill, North Carolina

Arnold J. Mandell, M.D.
Professor and Chairman
Department of Psychiatry
School of Medicine
University of California at San Diego
La Jolla, California

Bruce S. McEwen, Ph.D.
The Rockefeller University
New York, New York

James L. McGaugh
Professor and Chairman
Department of Psychobiology
University of California
Irvine, California

David S. Segal, Ph.D.
Department of Psychiatry
School of Medicine
University of California at San Diego
La Jolla, California

Larry R. Squire
Department of Psychiatry
School of Medicine
University of California at San Diego
La Jolla, California

Larry Stein, Ph.D.
Wyeth Laboratories
Philadelphia, Pennsylvania

Richard E. Whalen, Ph.D.
Professor of Psychobiology
University of California
Irvine, California

John E. Wilson, Ph.D.
Division of Chemical Neurobiology
Department of Biochemistry
The University of North Carolina
Chapel Hill, North Carolina

C. David Wise, Ph.D.
Wyeth Laboratories
Philadelphia, Pennsylvania

CONTENTS

xi

MEMORY

GONADAL HORMONES, THE NERVOUS SYSTEM AND BEHAVIOR

Richard E. Whalen

Department of Psychobiology, University of California

Irvine, California

Gonadal hormones have two major effects upon those systems which control sexual behavior. In the adult organism the gonadal hormones stimulate cells of fixed character and permit the display of sexual responses--these might be termed the concurrent or activational effects of hormones. In the developing organism, hormones act upon cells which are labile and determine their future responsiveness to hormones--these might be termed the differentiating effects of hormones. The present chapter will explore problems and raise the following questions with respect to the nature of the hormonal activation and differentiation of the substrate for behavior:

1. What? What are the effects of manipulating gonadal hormones upon behavior?
2. Which? Which hormones are necessary and/or sufficient in behavior control?
3. Where? Where do the hormones act?
4. How? What are the mechanisms of action by which gonadal hormones alter behavior?

As will become clear during this review each successive question becomes more and more difficult to answer with any precision.

EFFECTS OF HORMONES UPON BEHAVIOR

For our present discussion I will focus upon sexual behavior using the mating pattern of the laboratory rat as a model. The male and female rat display distinctive sets of sexual responses.

1

The male exhibits three major response patterns, mounting responses
without intromission, mounts with intromission or the insertion of
the penis into the vagina of the female, and ejaculation. Each
of these responses is behaviorally distinct. In the laboratory,
the rat will display these responses throughout the year primarily
during the dark phase of the diurnal cycle. The female rat also
displays distinctive sexual responses, but these occur only during
a limited period of 15-20 hours during a four or five day estrous
cycle. Female responses include the assumption of the lordosis
posture, a concave arching of the back which exposes the perineum,
ear-wiggling, and brief darting responses. The probability and
intensity of these responses waxes and wanes during the period of
heat.

For both male and female rat quantitative as well as qualita-
tive changes in behavior appear to reflect changes in sexual
excitability. For example, a male which has not mated for a month
will continue to engage in copulatory activity for three or more
hours during which time he will ejaculate 7-8 times. Two days
later he is unlikely to engage in any sexual activity, and if he
does so, he may obtain only one or two ejaculations before ceasing
to mate. In the male, the probability, latency and frequency of
the separate copulatory responses all appear to reflect his state
of motivation.

In the female the probability and quality of the lordosis
response appear to reflect her state of sexual receptivity. Early
during the heat period the female is likely to reject the advances
of the male. As time proceeds, the probability that a mount by a
male will elicit a lordosis response increases until at the point
of maximum heat nearly every mount by the male will elicit lordosis.

It is with these response patterns that we will be concerned
in our analysis of the effects of gonadal hormones upon behavior.

HORMONE REMOVAL AND REPLACEMENT

Removal

The effects of gonadectomy are rather different in adult male
and female rats. In the male, the removal of the testes is followed by a slow progressive decline in sexual capacity. The
latency to mount and achieve intromission increases as does the
interval between the first intromission and the occurrence of
ejaculation. (The sexually rested male rat achieves his first
ejaculation after approximately 11 intromissions, his second after
6, his third after four; subsequent ejaculations usually follow
3-5 intromissions). Following castration there is also a progressive increase in the duration of the post-ejaculatory intervals,

those periods of no mating which follow each ejaculation.

It has been reported that following castration ejaculation responses disappear first, followed successively by intromission responses and mounting responses. In studies employing short tests the complete cessation of mating has been seen to occur within two or three weeks. Davidson (1966), however, noted a progressive decrease lasting over 10 weeks in the proportion of his males which ejaculated each week. One male continued to ejaculate for 18 weeks after gonad removal. Davidson also found increases in intromission latency, ejaculation latency and in the duration of the post-ejaculatory intervals.

Thus, the display of mating responses by the male rat is critically dependent upon the presence of the testes. Yet a curious discrepancy seems to exist. Mating continues to occur for several weeks after castration, while gonadal hormones disappear from the circulation rather rapidly. To account for this apparent discrepancy it has been postulated that androgens from the adrenal gland moderate the decline in sexual activity. Bloch and Davidson (1968), however, have found no support for this hypothesis in their study of the effects of castration in adrenalectomized rats.

In the female rat, mating behavior disappears within hours of ovariectomy. The precise effects of ovariectomy nonetheless depend upon the exact time of operation within the estrous cycle. According to Schwartz (1969), if the ovaries are removed prior to onset of the surge of estrogen secretion during proestrus, mating will not occur as expected some 20 hours later. If the ovaries are removed 6 hours before expected mating, that is, following the estrogen surge, but prior to the normal progesterone surge, mating intensity is reduced, but the behavior is not eliminated (Powers, 1970). In the same study Powers also showed that ovariectomy 6 hours after the peak progesterone secretion had no effect upon the display of lordosis 6 hours later—86--96 per cent of the mounts by the males elicited lordosis. These animals failed to show lordosis 12 hours later, however, even if administered exogenous progesterone. Thus, the sexual responses of the female rat, unlike those of the male, rapidly disappear following gonadectomy.

Replacement

The identity of the gonadal hormones has been known for some years. The testes secrete both testosterone and androstenedione, the former being the predominant androgen both during development and in adulthood (Resko, Feder and Goy, 1968). In the rat, testosterone, but not androstenedione, is detectable in the blood at birth and five days after birth. Levels then drop until the animal is approximately 30 days of age (Resko et al., 1968). From

30 days of age onward plasma testosterone levels rise to a peak
which is reached between 60-90 days of age and then they fall
(Resko et al., 1968; Grota, 1971). Androstenedione becomes detectable at about 30 days of age and reaches its peak about 60 days of
age (Resko et al., 1968).

The ovaries of the female secrete five different hormones,
three estrogens, estradiol, estrone, and estriol, plus progesterone and 20 α - hydroxy-progesterone. Estradiol is considered to
be the most active estrogen and progesterone the more active progestin. Cyclic secretion of the hormones begins at puberty in the
female rat (approximately 35 days of age) although some evidence
would indicate that these hormones may be present before puberty.

While the gonads secrete their hormones in the free form,
much of what we know about hormone replacement has involved the
application of esters of the naturally occurring hormones such as
testosterone propionate (TP) and estradiol benzoate (EB). The
esterified forms of the hormones are more potent and have a greater
duration of action than the natural hormones and therefore have
been the agents of choice of most laboratories.

In 1949 Beach and Holz-Tucker reported on the efficacy of
testosterone propionate in maintaining mating behavior following
castration. Their animals were given six mating tests prior to
castration; daily therapy was started 48 hours after operation
and consisted of the administration of oil containing no hormone
or between 25 and 500 micrograms of TP for various groups. They
found that 50 µg/day maintained the latency of the mounting
response of the males at about prepoperative levels; doses below
this increased latencies while higher doses reduced latencies
below preoperative levels. While 50 µg/day maintained mounting
latency, the dose was not completely effective in maintaining
behavior at prepoerative levels. Some of the animals in this
group failed to achieve intromissions and to ejaculate on each
postoperative test. Between 75-100 µg/day of TP was found to be
necessary to maintain ejaculatory performance at preoperative
levels. Doses of 500 µg/day greatly reduced mount latency and
increased, relative to preoperative scores, the proportion of males
which ejaculated twice during a test period.

Some years later, Whalen, Beach and Kuehn (1961) treated with
TP rats which had been castrated and allowed to stop mating. The
dose needed to restore mating was approximately 400 µg/day, a
great deal more than was needed by Beach and Holz-Tucker to
maintain mating when treatment was started shortly after castration.
Davidson and his colleagues (1971) have investigated this phenomenon
systematically and have found that eight times the dose of TP is
needed to restore mating in rats castrated for two months than is

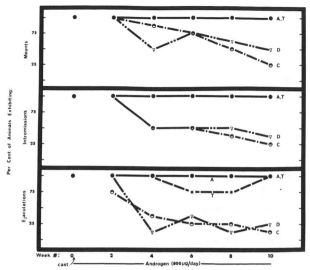

Figure 1. Proportion of male rats mounting, achieving intromission
and ejaculating prior to castration and following castration and
the administration of vehicle (C) or 800 µg/day testosterone (T),
androstenedione (A) or dihydrotestosterone (D). Data from Whalen
and Luttge, 1971a.

needed to maintain mating when hormone treatment is started at the
time of castration. Apparently continual hormonal stimulation is
needed if proper functioning of those cells which control behavior
is to be maintained.

While we do not understand the reason why more hormone is
needed to restore mating than is needed to maintain mating, what
is clear is the fact that TP is fully capable of stimulating all
systems involved in male behavior. The assumption that the active
agent is testosterone is not justified, however. Since the
important papers of Anderson and Liao and Bruchovsky and Wilson in
1968 we have known that the nuclei of the prostate, an androgen
dependent tissue, contains predominantly the 5 α reduced metabolite
of testosterone, dihydrotestosterone (5 α - androstane - 17 β - ol
- 3 - one). In a "maintenance" study we (Luttge and Whalen, 1970)
have found that this metabolite is somewhat more effective in
maintaining the weight of the seminal vesicles and prostate than
is testosterone itself. Thus the possibility existed that all
androgen sensitive tissues, including those involved with mating
behavior, respond not to the parent steroid, testosterone, but
rather to one of its metabolites. This, however, is not the case
as is illustrated in Figure 1 (Whalen and Luttge, 1971). Shown is
the proportion of rats mounting, achieving intromission and ejacu-

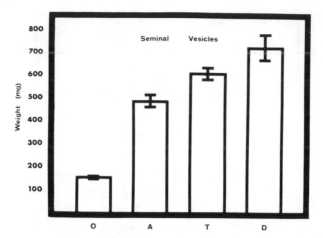

Figure 2. Mean seminal vesicle weight from male rats castrated
and administered oil (O) or 800 µg/day androstenedione (A), testosterone (T) or dihydrotestosterone (D) for 10 weeks. Data from
Whalen and Luttge, 1971a.

lation following castration and treatment with oil or 800 µg/day of
the free alcohol form of testosterone, androstenedione or dihydrotestosterone. Both testosterone and androstenedione maintained
mating at preoperative levels. The animals treated with dihydrotestosterone behaved as did the oil-treated castrates and showed a
progressive decline in mating performance. This figure also shows
that ejaculation disappeared first with mounts and intromissions
declining at approximately the same rate.

     Figure 2 shows that an inverse correlation exists between the
behavioral and somatic effects of these three steroids. Dihydrotestosterone, which had no effect upon behavior, was the most
effective steroid in maintaining seminal vesicle weight in these
animals. Similar findings have been obtained by McDonald et al.
(1970) and by Feder (1971).

     These data would thus indicate that male mating behavior in
the rat is under the control of either testosterone and/or androstenedione both of which are secreted by the testes.

     These findings may also help us to understand the curious fact
that male mating responses are not inhibited by the potent anti-
androgen, cyproterone acetate (Whalen and Edwards, 1969). This
agent can reduce the seminal vesicles almost to castrate size without influencing mating responses. Current evidence would suggest
that cyproterone acetate is particularly inhibitory with respect to

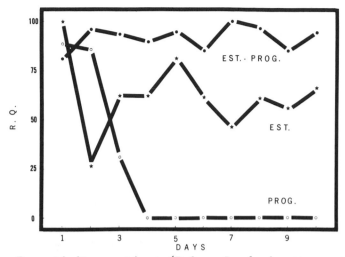

Figure 3. Receptivity quotient (R.Q. = Lordosis responses/mounts X 100) for ovariectomized female rats administered estradiol benzoate and progesterone and tested on Day 1 and then maintained on daily treatment with estradiol benzoate, estradiol benzoate plus progesterone, or progesterone alone. Data from Edwards, Whalen and Nadler, 1968.

dihydrotestosterone (Fang and Liao, 1969). If this proves to be the case, our assumption that behavioral systems are controlled by testosterone while peripheral tissue systems are controlled by dihydrotestosterone would be strengthened.

Mating behavior in the female rat can also be maintained and restored by the application of exogenous gonadal steroids. Figure 3 illustrates the effects of daily administration of estradiol benzoate, progesterone or both hormones upon the intensity of receptivity in ovariectomized female rats (Edwards, Whalen and Nadler, 1968). These data clearly demonstrate the well-known synergistic action of estrogen and progesterone (Beach, 1942). Estrogen is capable itself of inducing the display of lordosis, but when used without progesterone rather high doses of the hormone are required (Edwards et al., 1968; Davidson, Smith, Rodgers and Bloch, 1968; Whalen, Luttge and Gorzalka, 1971).

In this, and in most studies of the hormonal induction of receptivity, an esterified hormone such as estradiol benzoate was used. As I have mentioned earlier regarding the androgens, data from such studies cannot be taken as evidence that estradiol is the active agent. In a study of the localization of estrogenic metabolites in the brain of rats Luttge and I found that after the administration of tritiated estradiol both estradiol and estrone could be detected in hypothalamic tissue, relatively more estrone

and less estradiol being found in the posterior hypothalamus than
in the anterior hypothalamus (Luttge and Whalen, 1970). Similar
findings have been reported by Kato and Villee (1967). Moreover,
we have found opposite anterior to posterior gradients for the
neural accumulation of estradiol and estrone in brain tissue.
Estradiol concentrates predominantly in the anterior hypothalamus
while estrone concentrates in the posterior hypothalamus and
anterior mesencephalon (Luttge and Whalen, 1972). Thus it is
clear that estradiol is not the only ovarian steroid which could
act on brain tissue to control lordosis behavior.

Beyer, Morali and Vargas (1971) have recently shown that low
doses (1 µg/day/10 days) of estradiol and estrone, but not estriol
can increase sexual receptivity. At this dose estradiol was
significantly more effective than estrone. At a higher dose
(4 µg/day/10 days) animals in all treatment groups showed lordosis,
although again estriol was less effective than estradiol and
estrone although these two did not differ significantly.

The available data suggest therefore that estradiol is the
most potent estrogen with respect to sexual behavior in the rat,
although estrone may contribute to the control of sexual receptivity. One might wonder whether both estradiol and estrone are
effective because they act on different sets of cells as suggested
by the tracer studies mentioned earlier or whether the cells which
control lordosis are not entirely specific for specific estrogens.
The possible non-specificity of the system has gained some credence
by the repeated observation the it is possible to induce lordosis
in the rat by androgen treatment (Beach, 1942; Pfaff, 1970; Beyer
and Komisaruk, 1971). Data from our laboratory are shown in Figure
4 (Whalen and Hardy, 1970). Large doses of testosterone propionate
when combined with progesterone led to the display of lordosis
much as did estradiol benzoate treatment. When progesterone was
not administered TP failed to induce high levels of lordosis
behavior. A similar finding has been reported by Pfaff (1970).
Beyer and Komisaruk (1971), on the other hand, have shown that
free alcohol testosterone can be effective, even when progesterone
is not administered. They also reported that androstenedione is
ineffective in inducing lordosis under these conditions. We
(Whalen, Battie and Luttge, unpublished), however, have found that
androstenedione is effective if progesterone is also administered.
Thus, at least two androgens, testosterone and androstenedione,
are capable of inducing lordosis in female rats. Some authors have
suggested that these data indicate non-specificity for those neurons
which control behavior; others have suggested that the androgens
are metabolized to estrogens which are then the effective stimulating agents. The metabolic conversion of androgens to estrogens
has been demonstrated. The question whether sufficient estrogen
could be generated in this manner has remained open, however.

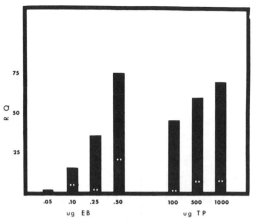

Figure 4.   Receptivity quotient (R.Q. = Lordosis responses/mounts X 100) for ovariectomized female rats administered estradiol benzo-ate or testosterone propionate daily in various doses. The animals were tested prior to (dashed bar) and three hours following (top of column) the administration of progesterone. Data from Whalen and Hardy, 1970.

In my laboratory Miss Cynthia Battie and I have recently attempted to answer this question by combining androgen and anti-estrogen treatments. For this work we selected the anti-estrogen CI-628 produced by Parke-Davis. Arai and Gorski (1968) had shown that this agent could block estradiol benzoate induced lordosis in the rat. We argued that if androgen-induced lordosis is the result of an androgen to estrogen metabolism, this lordosis behavior should be blocked by the anti-estrogen. Similarly, we argued that if androgen is inducing lordosis directly it should be uninfluenced by anti-estrogen. Our findings are shown in Figure 5. Lordosis was induced by TP and was not induced when CI-628 was administered at the same time and following TP injection. This inhibition was not permanent as the animals were capable of responding to TP on the following week.

Of course, it might be argued that the CI-628 was anti-andro-genic as well as anti-estrogenic. Two control studies suggested that this was not the case. In the first study we found that CI-628 would not antagonize seminal vesicle weight maintenance by TP given to castrated males. In the second study we found that CI-628 would not inhibit male mating responses maintained by TP. It is not like-ly, therefore, that CI-628 is a potent anti-androgen, yet it was very effective in inhibiting androgen-induced lordosis behavior. As a result of these findings we are led to believe that androgen-induced lordosis is a result of estrogen stimulation and that the system which controls lordosis is in fact specific to estrogens.

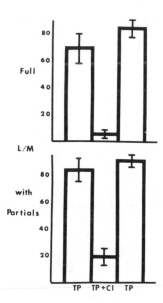

Figure 5.   Ratio of lordosis responses to mounts X 100 for ovariectomized female rats administered testosterone propionate plus
progesterone on week 1, testosterone propionate, progesterone and
the anti-estrogen CI-628 on week 2 and testosterone propionate
plus progesterone on week 3.   CI-628 is 1- [2- [p- [α-(p-Methoxy-
phenyl)-β-nitrostyryl] phenoxy] ethyl] pyrrolidine, monocitrate
(Parke-Davis).   Data from Whalen, Battie and Luttge, 1972.

     If we return to Figure 3 we are reminded that both estrogen
and progesterone play a role in the induction of sexual receptivity.   Progesterone, however, is not the only progestin secreted
by the rat.   In fact, 20 α - hydroxyprogesterone (20 α -
hydroxypregn - 4 - en - 3 - one) is secreted at higher levels
than is progesterone throughout the estrous cycle of the rat
(Talegdy and Endroczi, 1963; Uchida, Kadawski, and Miyake, 1969).
Zucker (1967) examined the relative effectiveness of these two
progestins and of pregnenolone and 17 α - hydroxyprogesterone.
Progesterone induced lordosis in estrogen pretreated rats,
pregnenolone was weakly effective and the other hormones were
almost completely without effect.   It seems reasonable to conclude
that progesterone is the primary steroid which synergizes with
estrogen to bring about lordosis.

     To this point we have asked two questions.   1. What are the
effects of gonadal hormones upon mating responses in the rat? and
2. Which hormones are involved?   A vast number of studies have
shown that gonadal hormones are necessary for the maintenance of
sexual behavior--the behavior disappears following gonadectomy

and can be restored by treatment with exogenous steroids. Accumulating data indicate further that both testosterone and androstenedione are capable of maintaining mating in male rats, although based upon secretion patterns, the former hormone is probably the active agent under natural conditions. In the female, both estradiol and estrone are capable of inducing lordosis at low doses. The active progestin would appear to be progesterone. Our next question is where do these hormones act to mediate behavioral responses.

SITE OF ACTION OF GONADAL HORMONES

Female Behavior

Since the early investigations of the hormonal control of sexual behavior in the late 1930s it has been our belief that this control is mediated by the actions of hormones upon the central nervous system. It was not until 1958, however, that evidence was forthcoming to support this hypothesis. At that time Harris, Michael and Scott reported that implants of synthetic estrogens into the diencephalon of cats could lead to the display of the mating posture typical of that species. These implants induced behavior while having no effects upon the peripheral estrogen-sensitive reproductive apparatus of the ovariectomized animals indicating that the hormones were not having their behavioral effects by being taken into the systemic circulation.

The first study employing the female rat was published by Lisk in 1962. This worker implanted either 27 ga. or 30 ga. tubes into the diencephalon of ovariectomized rats. The hormone was exposed only at the lumen of the tubes. These animals were tested daily with a male for the display of lordosis. Lordosis was found to occur in those animals with implants in the preoptic area, ventral to the preoptic area, in the anterior hypothalamic area and in the suprachiasmatic nucleus. Negative points were found in various posterior hypothalamic nuclei including the arcuate nucleus. Vaginal smears indicated that these females were diestrous at the time they were showing lordosis.

In later work Lisk (1966) placed similar implants into the region of the arcuate nucleus of females whose ovaries had not been removed. These implants altered the normal four day vaginal smear pattern. Nine of ten animals were diestrus yet mated every day. In still another set of animals the estrous cycle was made irregular and females mated without regard to their vaginal cycle. Lisk felt that these arcuate nucleus implants were effective because of a diffusion of hormones to the preoptic-suprachiasmatic region which his earlier research had indicated was the active site of estradiol action. Data from our laboratory are not con-

sistent with this interpretation, but rather suggest that an
extensive system of estrogen sensitive neurons exists in the brain
(Whalen and Hardy, unpublished).

We implanted 30 ga tubes containing estradiol or estradiol
benzoate into a variety of basal brain structures of the ovariectomized rat. We, as Lisk, found that implants in and about the
preoptic region increased receptivity. In addition, implants as
far anterior as the diagonal band of Broca and as far posterior as
the arcuate nucleus were effective. In our study we also examined
the effect of subcutaneously administered progesterone in animals
with central implants of estrogen. The details of the behavior of
two of these animals are presented in Table 1. In the animal with
a diagonal band implant of estradiol benzoate lordosis behavior
appeared on tests on which progesterone was not administered; the
probability of lordosis was increased by progesterone, however.
In the animal with arcuate nucleus implant of estradiol lordosis
behavior appeared only following progesterone stimulation. In both
cases the vaginal smear pattern was diestrus throughout testing and
the uterus was found to be atrophic at autopsy. These data are
consistent with those of Dorner, Docke and Moustafa (1968a) who
found that implants of 1 μg estradiol benzoate were effective
primarily when located in the ventromedial and arcuate nuclei.
Thus, a rather large area of the basomedial brain seems to respond
to estrogen in the control of the lordosis response.

Of course, lordosis is under the control not only of estrogen,
but of progesterone. Unfortunately, there has been rather little
study of those neural sites which respond to progesterone. Lisk
(1967) reported some preliminary work in which he implanted estradiol into the suprachiasmatic region on one side of the brain. The
rats mated briefly and then stopped. Progesterone was then implanted into the opposite suprachiasmatic region. Within 6 hrs.
four of the five rats displayed lordosis.

Recently, Ross, Claybaugh, Clemens and Gorski (1971) have
reported that mesencephalic rather than diencephalic implants of
progesterone facilitate lordosis in estrogen-primed rats. These
workers used a double-wall cannula system in which progesterone
could be applied to the brain of unanesthetized rats. The
probability of lordosis (lordosis responses/mounts by the male)
did not exceed 30% on the average when progesterone was placed in
the medial preoptic area, the lateral preoptic area, the anterior
hypothalamic area or in the ventromedial nucleus. However, when
the progesterone was placed into the mesencephalic reticular formation the probability of lordosis went to 50% 15 minutes after
implantation, to 80% within one hour and to nearly 90% within two
hours of implantation. The effective implants were located lateral
to the interpeduncular nucleus and dorsal to the pons.

TABLE 1

Effects of implants of estrogen in the central nervous system upon the display of lordosis. On each test day behavior was assessed before and three hours after the administration (SC) of 500 µg progesterone (from Whalen, R.E. and Hardy, D.F. unpublished).

Locus of Implant	Days Since Implant	Lordosis Responses/Mounts X 100	
		Before Progesterone	After Progesterone
Diagonal band of Broca (DeGroot 4.6)	5	6.7	64.3
	11	38.9	57.7
	15	33.3	56.2
	19	28.0	50.0
	26	66.7	91.7
	34	0	15.8
	42	28.6	66.7
	47	16.7	42.3
Basolateral arcuate nucleus (DeGroot 8.6)	7	0	57.1
	12	0	64.7
	15	0	77.3
	19	0	83.3
	22	0	60.0
	26	0	84.2
	29	10.0	71.4

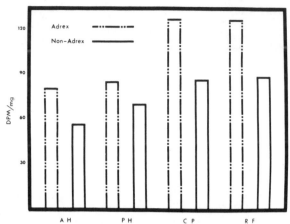

Figure 6. Radioactivity levels (disintegrations per min/mg) in
various brain areas one hour after the i.v. administration of
tritiated progesterone to ovariectomized female rats which were
either adrenalectomized or adrenally intact. AH = anterior
hypothalamus; PH = posterior hypothalamus; CP = cerebral peduncle
region of anterior mesencephalon; RF = mesencephalic reticular
formation. Data from Whalen and Luttge, 1971b.

     The findings of Ross and his colleagues were particularly
interesting to us because we have found a striking concentration of
radioactivity in the mesencephalon of female rats following the
administration of tritiated progesterone. Radioactivity concentrated in diencephalic structures as well, but not to the extent
found in the region of the cerebral peduncle of the anterior mesencephalon or in the reticular formation (Whalen and Luttge, 1971b,
1971c). These findings are illustrated in Figure 6. The possibility thus exists that the synergistic action of progesterone with
estrogen reflects the action of the different steroids at different
neural loci. Of course, these findings are rather new and deserve
to be replicated extensively before we speculate on the mechanism
by which estrogen and progesterone interact in the control of
lordosis behavior.

     Finally, for completeness, it should be mentioned that
Dorner, Docke and Moustafa (1968a) have found that implants of
testosterone propionate into the ventromedial-arcuate region lead
to the display of lordosis by ovariectomized female rats.

                          Male Behavior

     As with the female, controversy exists concerning the loca-
tion where steroid implants can activate masculine sexual responses.

In the first paper to appear using the rat as subject, Fisher (1956) reported that the injection of sodium testosterone sulfate in solution into the lateral preoptic area could lead to mating behavior shown with "exaggerated speed and compulsiveness." Unfortunately positive results were obtained in only six of 130 animals.

Davidson (1966) was able to induce male behavior in castrated rats much more reliably by applying crystalline testosterone pro- pionate to brain tissue. Thirty-nine of his rats showed copulation and ejaculation in the presence of atrophied accessory sex glands. The most effective loci included the medial preoptic region and a hypothalamic region caudal to the posterior hypothalamic nucleus but anterior to the habenulopeduncular tract. Implants in these areas not only maintained ejaculatory performance, but in some cases reduced initial intromission latency to preoperative levels.

Lisk (1967) had somewhat less success than Davidson. Only four of his animals showed ejaculation (6 others showed mounting); all of these animals had their implants situated in the anterior hypothalamic region. Unlike Davidson, Lisk found no positive loci in posterior hypothalamic areas.

Finally, there has been one report of male-like mounting behavior in female rats implanted with TP. Dorner, Docke and Moustafa (1968b) found the anterior hypothalamus, but not the posterior hypothalamus to be active in this regard.

The studies reviewed here must leave us dissatisfied with the current literature in what might be called "steroid implant neuronography." Rather few studies have been published, and those which have, have provided us with a rather incomplete picture of those neuronal systems which control masculine and feminine sexual responses. What does seem to have emerged, however, is the idea that the systems which respond to testosterone and estrogen are not discretely localized within any nuclear group of the diencephalon. Active sites have been found as far anterior as the diagonal band (estradiol) and preoptic area (testosterone) and as far posterior as the arcuate nucleus (estradiol) and mammillary nucleus (testosterone). This pattern of steroid sensitive sites is quite consistent to those revealed by autoradiographic and liquid scintillation techniques (Eisenfeld and Axelrod, 1966; Kato and Villee, 1967; Pfaff, 1968; Stumpf, 1968, 1970; Whalen, Luttge and Green, 1969; Luttge and Whalen, 1970; Whalen and Luttge, 1971; and others). What seem to exist are complex, anatomically extended systems which run throughout the basal brain. This revelation should make us very cautions of using concepts which suggest the existence of localized "sex centers."

Before leaving this topic we should broaden our concern with the site of action of gonadal hormones. Recent work by Hart (1967, 1969; Hart and Haugen, 1968) has directed our attention toward the action of hormones on the spinal cord. Hart has shown that the male rat is capable of displaying a variety of genital reflexes following cord transection. Erections, and what are termed "quick-flips" and "long flips" of the penis appear, and appear in response clusters. Following castration the frequency of these responses declines; precastration levels can be returned by the administration of TP. Hart and Haugen further showed that the frequency of these reflexes could be increased in castrated cord-transected males by implantation of testosterone propionate into the cord. Thus it would appear that some of the action of androgen on the genital reflexes is mediated directly by the spinal cord.

In contrast to the male, Hart has failed to find evidence that estrogen and progesterone can facilitate postural reflexes which resemble lordosis in the spinal female. The spinal female can display a lordosis-like response, but the frequency of this response is not increased by hormone administration.

Finally, we must recall the action of gonadal hormones upon the somatic tissues which are involved in mating, the penis of the male and vagina of the female in particular. The integrity of the morphology of the rat's penis, like its behavior, is controlled by androgens. In 1950 Beach and Levinson demonstrated that castration results in a progressive decline in the number of cornified papillae which are located on the surface of the glans penis. Five micrograms of TP daily slows, but does not prevent this decline; 75 µg/day, however, prevents the regressive changes which accompany castration. When Beach and Levinson compared the number of papillae with the frequency of ejaculation in castrates receiving various doses of hormone, the correlation was indeed striking. Thus the possibility existed that the hormones were exerting their action not on the brain, but rather by controlling the potential for sensory input through the penis. A number of studies have demonstrated that the integrity of the penis is important for the successful execution of copulatory responses (Whalen, 1968).

This almost invariable correlation between copulatory performance and penile morphology has been partially resolved by Beach and Westbrook (1968). These workers administered the synthetic androgen fluoxymesterone (9 α - fluro-11β - hydroxy - 17 - methyltestosterone) to castrated male rats. Five of the seven rats had ceased to copulate completely before androgen treatment. None of these animals copulated while under treatment. The two rats which continued sporadic copulatory activity after castration

did so while receiving fluoxymesterone, but at a level which was
not higher than before treatment. At sacrifice, the androgen
treated rats had 105 papillae per cross section of the glans penis,
intact males had 109, and untreated castrates had only 5. Thus,
it was possible to dissociate the behavior from penile morphology
indicating that a "normal" phallus is not a sufficient condition
for copulation to occur.

The reverse condition, however, has not been established,
viz., completely normal mating with an atrophic phallus. While
Davidson (1966) did show that male rats with brain implants of TP
will copulate and ejaculate with an atrophic phallus, his animals
did not perform at preoperative levels on all tests. The possibility remains that penile morphology must be within the normal
range if mating performance is to be normal.

With respect to the female and the need for a normal vaginal
state the evidence seems more convincing. Female rats with brain
implants of estrogen and an atrophic vagina can show a probability
of lordosis which is equivalent to that shown by intact females.
No one to date, however, has shown that other parameters of the
behavior, such as the intensity and duration of lordosis, or the
frequency of ear-wiggling and hopping, are restored to normal
levels. They may well not be, since Hardy and DeBold (1971) have
recently shown that even massive doses of estradiol benzoate
administered systemically fail to return lordosis duration to the
levels shown preoperatively.

Considering now all of the evidence concerning the site of
action of gonadal hormones in the control of sexual responses we
may conclude that the hormones work at several levels of the
biological system. Depending upon the sex of the animal and the
particular hormone involved, the hormones stimulate neurons
throughout the diencephalon, within the mesencephalon and in the
spinal cord and they stimulate the somatic tissues of the
genitalia.

SEXUAL DIFFERENTIATION

Since the important paper by Phoenix, Goy, Gerall and Young
in 1959 we have come to believe that gonadal hormones are involved
in the sexual differentiation of the brain. Phoenix et al. showed
that the female offspring of guinea pigs administered large doses
of TP during pregnancy are unlikely to display lordosis in adulthood even when administered exogenous estrogen and progesterone.
Similar findings were reported by Harris and Levine (1965) who
administered TP to newborn female rats. In the years following
these germinal experiments numerous investigators have asked questions about the role hormones play in the development of those

neural systems which control sexual behavior. A great deal of the
evidence relating to this topic has been reviewed recently by this
writer (Whalen, 1971) so my view of the state of our knowledge
will not be detailed here. A few comments seem appropriate.

We can ask the same questions of the hormonal control of the
development of sexuality as we have asked about the hormonal control of the sexual responses of adult animals, namely, what are
the effects of hormones, which hormones are involved, where and
how do they act. Table 2 summarizes evidence from many experiments
regarding the question of what are the effects of hormones upon
development. The probability of female behavior (lordosis) following estrogen and progesterone treatment, and the probability of
male-type behavior (mounts, intromissions and ejaculations) following testosterone treatment are indicated. As can be seen, both
males and females are very likely to show mounting responses in
adulthood regardless of their hormonal state during development.
Intromission and ejaculation responses are only seen in animals
hormonally stimulated shortly after birth. We have interpreted
these findings as indicating that gonadal hormones are not necessary for the development of a masculine sexual motivation system,
but only necessary for a masculine performance system. Fully
adequate performance seems to require the complete development of
a phallus and normal phallic development occurs only in hormone
stimulated animals. The correlation between phallic size and
intromission ability is striking (Beach, Noble and Orndoff, 1969).

With respect to female behavior, lordosis is seen in females
which develop with or without their ovaries suggesting that the
feminine response system is not enhanced by ovarian hormones, and
lordosis is seen in males which are castrated at birth. Normal
males and females, TP-treated at birth, fail to respond to estrogen
(Whalen, Luttge and Gorzalka, 1971) or to estrogen and progesterone
(Whalen and Edwards, 1967) in adulthood. Thus, testicular secretions or TP can serve to defeminize the organism.

Since TP seems to mimic the effect of testicular secretions,
it is generally assumed that defeminization is brought about by
testicular testosterone. Testosterone is present in the plasma of
the newborn rat (Resko, Feder and Goy, 1968). We have recently
obtained data which make us question this assumption however
(Luttge and Whalen, 1970). Instead of using the propionic ester of
testosterone, we administered the free-alcohol form of testosterone,
as well as dihydrotestosterone and androstenedione to newborn female
rats. Doses ranged up to 800 μg of steroid. None of these steroids,
not even testosterone, inhibited the animals' potential to show
lordosis in adulthood. All three steroids masculinized the genitalia, and testosterone and androstenedione produced anovulatory
sterility, indicating that the steroids were in fact active. Thus,

TABLE 2

Relationships between Hormonal Conditions during Development
and Sexual Responding of the Rat in Adulthood.

Hormonal condition during Critical Period	Hormonal Condition during Adulthood			
	Androgen			Estrogen and Progesterone
	Mounts	Intromission Responses	Ejaculation Responses	Lordosis Responses
Male				
1. Testes Intact	+ + +	+ + +	+ + +	—
2. Castrated at Birth	+ + +	+	—	+ + +
3. Castration at Birth + TP	+ + +	+ + +	+ +	—
Female				
4. Ovaries Intact	+ + +	+	—	+ + +
5. Ovariectomize at Birth	+ + +	+	—	+ + +
6. TP at Birth	+ + +	+ +	+	—
7. TP Pre- and Post-Natally	+ + +	+ + +	+ + +	—

we were not able to mimic the defeminizing action of TP with free
alcohol testosterone, the steroid which is secreted by the testis.
The simple conclusion that defeminization is induced by testicular
testosterone is not supported by our study. Of course, it is
possible that our dose was too low. On the other hand, the possibility exists, and must be considered, that the testes inhibit the
male's potential to show lordosis by the secretion of some unknown
substance other than testosterone. These studies must be continued
if we are to learn the nature of the agent which controls sexual
differentiation of the brain.

Next we may ask which neural structures are involved in differentiation. Here, unfortunately, there is very little evidence.
Nadler (1968) attempted to answer this question by implanting
minute amounts of TP into the brain of newborn rats. This worker
was able to alter the normal pattern of mating behavior and he
partially inhibited lordosis. The effective implants were in both
the hypothalamus and amygdala. There was, however, some evidence
for peripheral effects of these implants, on phallic growth for
example, so that the localization must be considered imprecise.

Beyond this we can say little about which neural areas are
altered when the rat is hormonally stimulated during the critical
period of sexual differentiation.

MECHANISMS OF HORMONE ACTION

To date we know rather little about how hormones influence
brain tissue in the control of sexual behavior. The information
which we do have is discussed by McEwen in another chapter in
this volume. I might, however, emphasize a few points. It would
appear that hormones are selectively accumulated by specific cells
within the central nervous system. The hormones pass through the
cell membrane to be bound by cytoplasmic macromolecules (Eisenfeld,
1970). The hormone is then transferred to the nucleus where it
remains in a bound form (Zigmond and McEwen, 1970). Presumably
the interaction of the hormone with the nuclear material sets in
motion a train of events which ultimately results in the display
of sexual behavior. While the nature of these events remains
unknown, it is likely that they involve changes in neural and
neurosecretory activity. A specification of these events remains
an important challenge for the coming decade.

ACKNOWLEDGMENT

Personal research was supported by grant HD-00893 from the
National Institute of Child Health and Human Development.

REFERENCES

Arai, Y. and Gorski, R. A. 1968. Effect of anti-estrogen on
  steroid induced sexual receptivity in ovariectomized rats.
  Physiol . Behav., 3, 351–353.
Beach, F. A. 1942. Male and female mating behavior in pre-puber-
  ally castrated female rats treated with androgens. Endocrinology,
  31, 673–678.
Beach, F. A. 1942. Importance of progesterone to induction of
  sexual receptivity in spayed female rats. Proc. Soc. Exptl.
  Biol. Med., 51, 369–371.
Beach, F. A. and Holz-Tucker, A. M. 1949. Effects of different
  concentrations of androgen upon sexual behavior in castrated
  male rats. J. comp. physiol. Psychol., 42, 433–453.
Beach, F. A. and Levinson, G. 1950. Effects of androgen on the
  glans penis and mating behavior of castrated male rats. J. exp.
  Zool., 114, 159–171.
Beach, F. A. and Westbrook, W. H. 1968. Dissociation of androgenic effects on sexual morphology and behavior in rats.
  Endocrinology, 83, 395–398.
Beach, F. A., Noble, R. G., and Orndoff, R. K. 1969. Effects of
  perinatal androgen treatment on responses of male rats to
  gonadal hormones in adulthood. J. comp. physiol. Psychol., 68,
  490–497.
Beyer, C. and Komisaruk, B. 1971. Effects of diverse androgens
  on estrous behavior, lordosis reflex, and genital tract morphology in the rat. Horm. Behav., 2, 217–226.
Beyer, C., Morali, G., and Vargas, R. 1971. Effect of diverse
  estrogens on estrous behavior and genital tract development in
  ovariectomized rats. Horm. Behav., 2, 273–277.
Bloch, G. J. and Davidson, J. M. 1968. Effects of adrenalectomy
  and experience on postcastration sex behavior in the male rat.
  Physiol. Behav., 3, 461–465.
Davidson, J. M. 1966. Activation of the male rat's sexual
  behavior by intracerebral implantation of androgen. Endocrinology,
  79, 783–794.
Davidson, J. M. 1966. Characteristics of sex behaviour in male
  rats following castration. Anim. Behav., 14, 266–272.
Davidson, J. M. 1971. Hormones and reproductive behavior. In
  Reproductive Biology, H. Balin and S. Glassen (Eds.). Excerpta
  Medica.
Davidson, J. M., Smith, E. R., Rodgers, C. H., and Bloch, G. J.
  1968. Relative thresholds of behavioral and somatic responses
  to estrogen. Physiol. Behav., 3, 227–229.
DeGroot, J. 1959. The rat forebrain in stereotaxic coordinates.
  N.V. Noord-Hollandische Uitgevers Maatschoppij, Amsterdam.
Dorner, G., Docke, F., and Moustafa, S. 1968. Homosexuality in
  female rats following testosterone implantation in the anterior
  hypothalamus. J. Reprod. Fert., 17, 173–175 (b).

Dorner, G., Docke, F., and Moustafa, S. 1968. Differential
  localization of a male and a female hypothalamic mating centre.
  J. Reprod. Fert., 17, 583-586.
Edwards, D. A., Whalen, R. E., and Nadler, R. D. 1968. Induction
  of estrous: Estrogen-progesterone interactions. Physiol. Behav.,
  3, 29-33.
Eisenfeld, A. J. 1970. ^{3}H-estradiol: In vitro binding to
  macromolecules from the rat hypothalamus anterior pituitary and
  uterus. Endocrinology, 86, 1313-1318.
Eisenfeld, A. J. and Axelrod, J. 1966. Effect of steroid hormones
  ovariectomy, estrogen pretreatment, sex and immaturity on the
  distribution of ^{3}H-estradiol. Endocrinology, 79, 38-42.
Fang, S. and Liao, S. 1969. Antagonistic action of anti-androgens
  on the formation of a specific dihydrotestosterone-receptor
  protein complex in rat ventral prostate. Mol. Pharmacol., 5,
  420-431.
Feder, H. H. The comparative actions of testosterone propionate
  and 5αandrostan 17β-ol-3-one propionate on the reproductive behavior physiology and morphology of male rats. J. Endocrin., in
  press.
Fisher, A. E. 1956. Maternal and sexual behavior induced by
  intracranial chemical stimulation. Science, 124, 228-229.
Grota, L. J. 1971. Effects of age and experience on plasma
  testosterone. Neuroendocrinology, 8, 136-143.
Hardy, D. F. and DeBold, J. F. 1971. The relationship between
  levels of exogenous hormones and the display of lordosis by the
  female rat. Horm. Behav., 2, 287-297.
Harris, G. W., Michael, R. P., and Scott, P. P. 1958. Neurological site of action of stilbestrol in eliciting sexual behavior.
  In Ciba Foundation Symposium: Neurological Basis of Behavior,
  Churchill: London, pp. 236-254.
Harris, G. W. and Levine, S. 1965. Sexual differentiation of the
  brain and its experimental control. J. Physiol. (London), 181,
  379-400.
Hart, B. L. 1967. Testosterone regulation of sexual reflexes in
  spinal male rats. Science, 155, 1283-1284.
Hart, B. L. 1969. Gonadal hormones and sexual reflexes in the
  female rat. Horm. Behav., 1, 65-71.
Hart, B. L. and Haugen, C. M. 1968. Activation of sexual reflexes
  in male rats by spinal implantation of testosterone. Physiol.
  Behav., 3, 735-738.
Kato, J. and Villee, C. A. 1967. Preferential uptake of estradiol
  by the anterior hypothalamus of the rat. Endocrinology, 80,
  567-575.
Lisk, R. D. 1962. Diencephalic placement of estradiol and sexual
  receptivity in the female rat. Am. J. Physiol., 203, 493-496.
Lisk, R. D. 1966. Hormonal implants in the central nervous
  system and behavioral receptivity in the female rat. In Brain
  and Behavior: The Brain and Gonadal Function, R. A. Gorski and

R. E. Whalen (Eds.) UCLA Forum in Medical Sciences, No. 3,
  University of California Press: Los Angeles, pp. 98-117.
Lisk, R. D. 1967. Sexual behavior: Hormonal control. In
  Neuroendocrinology (Vol. 2), L. Martini and W. F. Ganong (Eds.)
  Academic Press: New York, pp. 197-239.
Lisk, R. D. 1967. Neural localization for androgen activation of
  copulatory behavior in the male rat. Endocrinology, 80, 754-761.
Luttge, W. G. and Whalen, R. E. 1970. Regional localization of
  estrogenic metabolites in the brain of male and female rats.
  Steroids, 15, 605-612.
Luttge, W. G. and Whalen, R. E. 1970. Dihydrotestosterone, andro-
  stenedione, testosterone: Comparative effectiveness in masculin-
  izing and defeminizing reproductive systems in male and female
  rats. Horm. Behav., 1, 265-281.
Luttge, W. G. and Whalen, R. E. 1972. The accumulation, retention
  and interaction of oestradiol and oestrone in central neural and
  peripheral tissues of gonadectomized female rats. J. Endocrin.,
  in press.
McDonald, P., Beyer, C., Newton, F., Brien, B., Baker, R., Tan, H.S.,
  Sampson, C., Kitching, P., Greenhill, R., and Pritchard, D. 1970.
  Failure of 5α-dihydrotestosterone to initiate sexual behavior in
  the castrated male rat. Nature (London), 227, 964-965.
Nadler, R. D. 1968. Masculinization of female rats by intracra-
  nial implantation of androgen in infancy. J. comp. physiol.
  Psychol., 66, 157-167.
Pfaff, D. W. 1968. Autoradiographic localization of radioactivity
  in rat brain after injection of tritiated sex hormones. Science,
  161, 1355-1356.
Pfaff, D. 1970. Nature of sex hormone effects on rat sex beha-
  vior: Specificity of effects and individual patterns of response.
  J. comp. physiol. Psychol., 73, 349-358.
Phoenix, C. H., Goy, R. W., Gerall, A. A., and Young, W. C. 1959.
  Organizing action of prenatally administered testosterone pro-
  pionate on the tissues mediating mating behavior in the female
  guinea pig. Endocrinology, 65, 369-382.
Powers, J. B. 1970. Hormonal control of sexual receptivity during
  the estrous cycle of the rat. Physiol. Behav., 5, 831-835.
Resko, J. A., Feder, H. H. and Goy, R. W. 1968. Androgen concentrations in plasma and testis of developing rats. J. Endocrin.,
  40, 485-491.
Ross, J., Claybaugh, C., Clemens, L. G., and Gorski, R. A. 1971.
  Short latency induction of estrous behavior with intracerebral
  gonadal hormones in ovariectomized rats. Endocrinology, 89, 32-
  38.
Schwartz, N. B. 1969. A model for the regulation of ovulation in
  the rat. Rec. Prog. Horm. Res., 25, 1-55.
Stumpf, W. E. 1968. Estradiol-concentrating neurons: Topography
  in the hypothalamus by dry-mount autoradiography. Science, 162,
  1001-1003.

Stumpf, W. E. 1970. Estrogen-neurons and estrogen-neuron systems in the periventricular brain. Am. J. Anat. 129, 207-218.

Telegdy, G. and Endroczi, E. 1963. The ovarian secretion of progesterone and 20α-hydroxypregn-4-en-3-one in rats during the estrous cycle. Steroids, 2, 119-123.

Uchida, K., Kadowski, M., and Miyake, T. 1969. Ovarian secretion of progesterone and 20α-hydroxypregn-4-en-3-one during rat estrous cycle in chronological relation to pituitary release of lutinizing hormone. Endocrinol. Japan, 16, 227-237.

Whalen, R. E. 1968. Differentiation of the neural mechanisms which control gonadotropin secretion and sexual behavior. In Perspectives in Reproduction and Sexual Behavior, M. Diamond (Ed.) Indiana University Press: Bloomington, pp. 303-340.

Whalen, R. E. 1971. The ontogeny of sexuality. In Ontogeny of Vertebrate Behavior, H. Moltz (Ed.), Academic Press: New York, pp. 229-261.

Whalen, R. E., Battie, C., and Luttge, W. G. 1972. Anti-estrogen inhibition of androgen induced sexual receptivity in rats. Behav. Biol., in press.

Whalen, R. E., Beach, F. A., and Kuehn, R. E. 1966. Effects of exogenous androgen on sexually responsive and unresponsive male rats. Endocrinology, 69, 373-380.

Whalen, R. E. and Edwards, D. A. 1967. Hormonal determinants of the development of masculine and feminine behavior in male and female rats. Anat. Rec., 157, 173-180.

Whalen, R. E. and Edwards, D. A. 1969. Effects of the anti-androgen cyproterone acetate on mating behavior and seminal vesicle tissue in male rats. Endocrinology, 84, 155-156.

Whalen, R. E., Luttge, W. G. and Green, R. 1969. Effects of the anti-androgen cyproterone acetate on the uptake of 1, 2 –^{3}H-testosterone in neural and peripheral tissues of the castrate male rat. Endocrinology, 84, 217-222.

Whalen, R. E. and Hardy, D. F. 1970. Induction of receptivity in female rats and cats with estrogen and testosterone. Physiol. Behav., 5, 529-533.

Whalen, R. E., Luttge, W. G., and Gorzalka, B. B. 1971. Neonatal androgenization and the development of estrogen responsivity in male and female rats. Horm. Behav., 2, 83-90.

Whalen, R. E. and Luttge, W. G. 1971. Testosterone, androstenedione and dihydrotestosterone: Effects on mating behavior of male rats. Horm. Behav., 2, 117-125 (a).

Whalen, R. E. and Luttge, W. G. 1971. Role of the adrenal in the preferential accumulation of progestin by mesencephalic structures. Steroids, 18, 141-146 (b).

Whalen, R. E. and Luttge, W. G. 1971. Differential localization of progesterone uptake in brain: Role of sex, estrogen pretreatment and adrenalectomy. Brain Res., 33, 147-155 (c).

Zigmond, R. E. and McEwen, B. S. 1970. Selective retention of oestradiol by cell nuclei in specific brain regions of the

ovariectomized rat. J. Neurochem., 17, 889–899.
Zucker, I. 1967. Actions of progesterone in the control of sexual
   receptivity of the spayed female rat. J. comp. physiol. Psychol.,
   63, 313–316.

NEUROENDOCRINE AND NEUROANATOMIC CORRELATES OF ATYPICAL SEXUALITY*

Richard Green, M.D.

Department of Psychiatry, University of California

School of Medicine, Los Angeles, California

NEUROANATOMIC ABNORMALITIES AND ATYPICAL SEXUAL BEHAVIOR

The presence within the same patient of both a demonstrable brain abnormality and unusual sexual behavior is provocative and tempts one to causally relate the two. For example, there have been several case reports of persons with brain tumors, usually temporal lobe in location in association with transvestism or other fetishism. Unquestionably, space occupying lesions can cause personality change as can dysrhythmias of neuronal discharge. Furthermore, from animal experimentation, it is clear that there are discrete areas within the brain responsible for some aspects of sexual behavior. Thus it is possible that in humans a lesion could, if properly located, have an effect on sexual behavior. However, magnitude of sexuality (hyper- and hyposexuality) would be the responses predicted. The occasional emergence of atypical sexual behavior, atypical in the direction of sexuality is of especial interest. Such occurrences have been viewed as evidence for disordered brain physiology, rather than psychodynamic factors, as being responsible for atypical sexuality.

Regretably, detailed clinical histories of patients with atypical sexuality and brain pathology are usually missing from case reports. Most often it is reported: "The patient showed no evidence of transvestism prior to the onset of his cerebral pathology at age 35." Researchers would be on more solid ground if details were known of cross-dressing activities in childhood, and the extent of any undue interest in cross-dressing in subsequent years, even if not practiced. Implications differ considerably whether transvestism with concurrent pathology becomes manifest in a person with no

27

previously latent impulses to cross-dress, or in someone with a
long-standing preoccupation, suddenly rendered less capable of suppression. Furthermore, it is also known that adults with no previous history of atypical sexuality may begin manifesting such behavior relatively late in life without evidence of anatomic brain
pathology. There may be many non-specific disinhibitors releasing
behavior, but a specific primary trigger of atypical sexual behavior would be of striking significance.

With increased sophistication in electroencephalographic techniques, interest has focussed on abnormal rhythms in association
with sexual variations. Particular attention has been paid to
transvestism and, less often, to other fetishism. Dysrhythmias
are of more theoretic and heuristic interest than space-occupying
lesions because they may represent subtler evidence of brain pathology, perhaps unrecognized without a specialized technique, and
possibly present in populations of seemingly healthy persons.
Furthermore, the disorder is more accessible to experimental manipulation by the use of anticonvulsant drugs which may modify electrical discharges and restricted surgical removal of abnormally-
discharging nerve cells.

An oft-quoted case report is that of a safety-pin fetishist
with temporal lobe epilepsy (Mitchell et al., 1954). In this case
a "perverse form of erotic gratification, the contemplation of a
safety pin, became attached to the onset of the epileptic seizure
. . . ." Surgical removal of the left anterior temporal lobe relieved both the epilepsy and sexual arousal to safety pins.

The case report of a thirty-eight year old woman with "sexual
seizures" in association with destruction of one temporal lobe is
of interest both for the vivid detail of the "automatic" sexual
behavior and the subsequent amnesia, characteristic of temporal
lobe epilepsy (Freeman and Nevis, 1969). The patient had primary
syphilis at 16. At 36 she complained of a four-year history of
feeling of itching in the pelvic area, accompanied by a feeling
that a red hot poker was being inserted into her vagina. The
patient would then spread her legs apart, beat both hands on her
chest, and "verbalize her sexual needs (often in vulgar terms)."
She would have no memory for these episodes. A pneumoencephalogram
revealed bilaterally dilated ventricles and a brain arterial study
showed changes consistent with tissue loss in the right temporal
lobe. This was thought to be due to central nervous system syphilis. An anti-convulsant drug brought the seizures under some degree of control.

Removal of specific brain areas resulting in changes in sexual
behavior is a long-standing observation. Over thirty years ago,
Kluver and Bucy (1939) described, in the male monkey, a behavior

pattern produced by bilateral removal of the temporal lobes with
the uncus, amygdala, and hippocampus. These monkeys displayed,
after 3-6 weeks, hyper-sexuality, with considerable masturbation
and mounting of males or females. In the human, a case has been
reported (Terzian and Dalle Ore, 1955) of a 19-year-old male who
had a similar operative procedure and two weeks later showed atypical sexual behavior. He reported his attention was attracted
by the sexual organs of an anatomic diagram hanging on the wall
and displayed to his doctor that he had spontaneous erections followed by masturbation and orgasm. He also showed, after surgery,
heterosexual indifference, in contrast to his previous behavior,
and made homosexual invitations. This changed behavior was
reported to have persisted for at least two years.

Temporal lobe function in transsexuals has been studied in
15 patients by Blumer (1969). Two patients had EEG records considered abnormal. Three other records, though not unremarkable, were
still considered within normal limits, while ten others were entirely unremarkable. A previous report (Walinder, 1965) had described
nearly half of 26 subjects who were transvestites, or probably
transsexuals, as having abnormal EEGs. However, no matched controls were included in that series, and it is possible that the
criteria of abnormality were not the same as in the study by Blumer.

A larger survey looked at the medical records of 86 men
treated in an anti-epileptic clinic in Czechoslovakia (Kolarsky
et al., 1967). A medical history indicating an event which may
have precipitated brain damage (usually infection or trauma) was
found more often among epileptics who also manifested a sexual
variation than among epileptics with typical sexuality. Furthermore, when comparing the two groups, those with atypical sexuality
had developed epilepsy earlier in life. However, since only 10%
of the epileptic men showed any unusual sexual behavior, this
percent may be no higher than the general non-epileptic population.

A man showing sexual arousal to hair who also had a left temporal lobe tumor has been described (Ball, 1968). As a child he
had grossly retarded motor development and a minor degree of hydro-
cephalus. Since age 4 he experienced an intense preoccupation with
women's long hair. From 10 on he had intermittently dressed in
women's clothes and, in his 20's, would pay prostitutes to allow
him to stroke their hair while being masturbated. Psychologic
treatment (aversion therapy) for the hair fetish resulted in loss
of symptoms for 18 months. Subsequently, major seizures ensued
during sleep and the hair fetish returned. Though his electroen-
cephalogram was still normal, he was found three months later to
have a brain tumor. Both the seizures and the hair fetish came
under some degree of control with anti-epileptic medication.

Neuroanatomical knowledge gained from animal research has been recently applied to treatment of atypical human sexuality (Roeder and Muller, 1969). In the rat, and other species, the role of the ventromedial nucleus of the hypothalamus in regulating gonadal hormone secretion has been elucidated. Extending the implications of this finding to the human, a team of surgeons elected to treat a 40-year-old male attracted to young boys by destroying that nucleus of the patient's brain in his non-dominant hemisphere. A seven year postoperative followup report indicated a reduction in sexual drive and potency and an absence of previous sexual tendencies. Urinary hormone levels and seminal fluid were described as normal.

A second patient treated more recently by the same team was sexually attracted to young adolescent males and had an aversion to females. He regarded his behavior as an organic disease and "at once agreed to a stereotaxic procedure to remove the 'sex-behavior centre.'" At short-term follow-up (6 months) he reported no homosexual fantasies and no further revulsion to women.

The third patient was an elderly male also sexually attracted to young boys. One year after neurosurgery he reported a sex drive diminished in intensity but not direction.

The extent to which the (apparent) effectiveness of this procedure is due to interference with the central regulation of androgen secretion via gonadotropins with its resultant loss of sex drive, or to a direct destructive influence of a hormone sensitive brain area, or to the high motivation for change and expectation of help by patients who agree to such a procedure cannot be assessed at this time. Also, the fact that one-sided brain lesions in animals do not appear to affect sexual behavior makes interpretation of this report difficult.

NEUROENDOCRINE ABNORMALITIES AND ATYPICAL SEXUAL BEHAVIOR

Recent studies have also focussed on the role neuroendocrine factors may play in psychosexual development. Sex hormones may differentiate the central nervous system in a manner analogous to that in the peripheral reproductive system (Grady, Phoenix and Young, 1965). Of considerable significance for understanding sexual development has been the finding that the basic biologic disposition of mammalian embryos is female. No gonads and no gonadal hormones are required for an organism to develop along female lines. For maleness to emerge, however, androgenic hormones must act at critical developmental periods. This was demonstrated initially in the rabbit when a male fetus was castrated in utero: subsequent anatomic development progressed along female lines (Jost, 1947). In the human, the syndromes of Turner (gonadal dysgenesis) and testi-

cular feminization (androgen insensitivity) strikingly illustrate
the analogous phenomenon. Children with Turner's syndrome generally
have but one sex chromosome (X), develop no gonads, and appear to
be female at birth (Money, 1968). Children with testicular feminization are chromosomally male (XY), and have testes which secrete
normal amounts of the male hormone, testosterone, but their body
cells are unable to utilize it (Simmer, Pion and Dignam, 1965;
Rivarola et al., 1967). At birth they appear to be normal females
(Money, 1968).

It is recently apparent that critical developmental periods
exist during which levels of androgenic hormone also influence postnatal sex-related _behavior_. For example, a "tomboy" female rhesus
monkey results when as a fetus she is exposed to unusually large
amounts of male hormone, in consequence of her mother being injected
with testosterone propionate. Normal preadolescent male and female
rhesus monkeys behave quite differently, much in the same was as do
human boys and girls. Thus, the male monkey more often partakes in
rough-and-tumble play, chasing activity, and is more threatening.
However, females who received male hormone _before_ birth, in addition to being anatomically virilized, are considerably more "masculine" in their behavior. As young females, they participate in
rough-and-tumble activity in a manner similar to males. Comparable
amounts of male hormone given _after_ birth, however, do not appear
to have the same masculinizing effect (Young et al., 1964).

These primate studies provide speculative appeal for a related
phenomenon operating in man. While it is not possible to conduct
parallel experiments with humans, there are some clinical circumstances in which human females have been exposed to unusually high
levels of male hormone before birth. These are girls with the
adrenogenital syndrome and the girls whose mothers received androgenic progestins during gestation to ward off abortion.

In the adrenogenital syndrome an enzymatic defect in the production of some adrenocortical hormones results in excessive production of other adrenal hormones which are genitally masculinizing.
This overproduction begins before birth and continues postnatally
unless treated.

One study has focussed on 15 girls exposed to excessive androgen before birth, but not after. The diagnosis of adrenogenital
syndrome had been made in infancy so that androgen excess was terminated. This natural experiment is somewhat analogous to the
monkey procedure mentioned earlier. These girls as preadolescents,
were compared with 15 girls in whom there was no evidence of male
hormonal excess before birth. When compared for their interest in
doll play, 8 of the normal girls enjoyed playing with dolls compared to only 2 of the androgenized girls. With respect to inter-

est in boys' toys, only one of the normal girls expressed such
interest compared to 8 of the androgenized girls. Eleven of the
androgenized girls were considered tomboys compared to none of the
non-androgenized females. Finally, with respect to whether the
child preferred being a boy or a girl, 14 of the controls preferred
being a girl compared to only 7 of the androgenized girls (Ehrhardt,
Epstein and Money, 1968). However, the investigators point out that
7 of the 15 androgenized girls had been thought to be boys at birth
but were reassigned as girls before 7 months. The parents knew
of the genital masculinization at birth such that "This knowledge
may have insidiously influenced their expectancies and reactions
regarding the child's behavioral development."

Also studied were a group of 23 adult females who were exposed
to excessive androgen levels not only _prenatally_ but because they
were not treated during childhood, for at least 8 years _after_ birth
as well. Sexual preferences as adults were assessed. Of the 23,
11 had had exclusively heterosexual relations and none were exclusively homosexual. Only 2 had had frequent homosexual relations.
There were no female-to-male transsexuals in the group (Ehrhardt,
Evers and Money, 1968).

A third group consisted of 10 girls age 3-14, exposed to masculinizing hormones prenatally, in consequence of their pregnant
mothers having received androgenic progestins to prevent abortion.
Progestins, although a "female" hormone in that they are required
for continuance of pregnancy, have a masculinizing effect.

On the "Draw-A-Person" test, 7 girls drew a female first, a
response believed to be an indicator of a female gender identity.
However, nine showed a strong interest in boys' toys, 6 showed an
interest in organized team sports, and 9 liked to compete with
boys in sports. Only 2 liked frilly dresses. Finally, 9 were called
"tomboy" by either their parents, themselves, or both. However, in
one family a sister who had _not_ received progestin was at least as
tomboyish as her progestin-exposed sister (Ehrhardt and Money, 1967).

The testicular feminizing syndrome is an experiment of nature
and a medical parallel to laboratory studies in which male fetuses
are deprived of androgen. These persons, unable to utilize male
hormone have undescended testes and the male chromosome pattern,
XY. At birth since they appear to be normal infant females, they
are raised as girls and later show appropriately feminine behavior.
At puberty, they develop feminine breasts, presumably via the influence of estrogens normally secreted by the testes of all males, but
here not countered by the effect of male hormone. The absence of
menstruation (there is no uterus) frequently leads to the diagnosis.
Such persons are very feminine, are not told they are chromosomal
and gonadal males and live their lives as women. It is possible

that the absence of a male-hormonal influence on the fetal nervous system enhances their capacity to so readily adjust to the female role (Money, Ehrhardt and Masica, 1968).

A neuroendocrine basis for transsexualism, where sex-change surgery is desired, is a provocative concept. From animal work it is evident that at least in some species there exists a period of behavioral sexual differentiation in response to male hormone exposure, as well as a period of genital differentiation, and that these two critical time periods may be separate (Whalen, Peck and LoPiccolo, 1966). Thus it is possible to approach, in the laboratory, a model of transsexualism in which a "female mind exists in a male body," and vice versa. This could result from an androgen deficiency, at one of the critical developmental periods, resulting in an anatomically normal-appearing male with an unmasculinized or undifferentiated central nervous system.

Rather than postulating a global neural organization by hormones in a male or female direction, the effect of an excess or deficiency of androgen could be on non-specific variables such as aggressivity and activity. These could be factors which subtly influence early mother-child and peer-child relations. For example, a passive boy might be treated more delicately by his parents and might find the games and companionship of girls more agreeable than the rough-and-tumble of more aggressive boyhood.

In the great majority of cases of transsexualism there is presently no evidence that such a hormonal imbalance may have existed. However, there are a few patients in whom there is some basis to make this speculation. Recently, there has been reported a series of males desirous of living as women in whom a testicular defect was discovered (Baker and Stoller, 1968).

Case One appeared to be a normal male at birth and was so raised. However, he developed a feminine social orientation, and behaved as a girl from age 4. At 28, after requesting surgical sex reassignment to live as a female, a testicular biopsy revealed an abnormality. The diagnosis of "Sertoli-Cell-Only Syndrome" was made. There is uncertainty as to whether such testes produce abnormal amounts of estrogen; however, it is of interest that Sertoli-Cell tumors in dogs are feminizing.

Case Two also appeared to be a normal male at birth but insisted during childhood on behaving as a girl. During adolescence his body became feminized with small, feminine breasts and well-developed nipples. Facial hair did not appear until his 20's. At 30, the patient requested surgical sex reassignment to live as a female. Testicular biopsy and chromosomal study revealed an intersexed chromosomal configuration, XXY, and small, hypofunctional testes.

Case Three, also born an apparently normal male, was also femininely-oriented from early childhood. At 27, he began living as a woman. Medical examination here too revealed underdeveloped testes, and a diagnosis of chromatin negative Klinefelter's Syndrome was made. Case Four, similarly feminine during childhood, was subsequently diagnosed as having a pituitary gland deficiency with one consequence being a secondary deficit in testicular output. As an adult he reported feeling like a woman unless his usual low levels of androgen were supplemented by injections of testosterone (It was not possible to rule out the effects of suggestion here on enhanced feelings of masculinity).

These cases, all with evidence of deficiently functioning testes, may represent the clinical result of male hormone deficiency at a critical period in the neural development of gender behavior in man. Or, they may represent the combined effect of specific experiential childhood factors superimposed on a receptive neural substrate, the latter influenced by androgen deficiency. Or, they may represent the coincidental existence of two independent phenomena, cross-gender identification and hypogonadism.

Since this latter report, another sample of males desirous of sex change has been studied using sophisticated biochemical measures. Both the pattern of gonadotropin secretion and target tissue responsivity to male hormone have been measured. Earlier studies of rodents indicated that a male, deprived of androgen at a critical developmental period, released gonadotropins from the hypothalamic-pituitary axis in a cyclic pattern similar to the normal female, rather than in the normal tonic male pattern (e.g., Harris, 1964). It has been demonstrated that the pattern is determined by androgen action on the hypothalamus. An indirect way of assessing whether there may have been a mid-brain androgen deficit during the development of male transsexuals thus presented itself: determining the pattern of gonadotropin release. Analysis of daily levels of ICSH (a gonadotropin) revealed the pattern to be tonic, i.e., normal male. A second strategy was employed which determined whether tissues normally highly receptive to the action of male hormone respond to a normal degree in male transsexuals. Scrotal skin of transsexuals and non-transsexuals was incubated with radioactively labelled androgen and the quantity of testosterone and its metabolites present in the tissues assessed. Here again, no differences between the two groups were found (Gillespie, 1971). However, since the peripheral reproductive organs of male-to-female transsexuals are typically normal, there is little reason to suspect a deficiency in androgen metabolism at such a site. These studies complement the earlier finding of Migeon and co-workers (1969) that the concentration of plasma testosterone and its metabolites as well as urinary estrogens was the same in a sample of male transsexuals and non-transsexuals. When coupled with the finding of Ehrhardt and Money that females with the adrenogenital syndrome do not become female-to-male transsexuals, and Jones' finding (1971) that a sample of female transsexuals had normal plasma testosterone levels, these are important, though not conclusive, negative findings to be considered by those who would ascribe transsexualism to a purely endocrine etiology.

A recent revival of interest in a hormonal basis of homosexuality has resulted from four endocrine studies comparing homosexuals and heterosexuals. In the first, 24-hour urine samples from 40 male subjects were analyzed for levels of the stereoisomers, androsterone and etiocholanalone, two metabolic products of testosterone. The ratio of urinary etiocholanalone vs androsterone differed for the homosexuals and heterosexuals. However, considerable caution must be exercised prior to concluding that the different ratio is directly related to sexual preference. Three heterosexuals reported in the same study, who were severely depressed, also had urinary levels like that of homosexuals, as did one heterosexual diabetic (Margolese, 1970). In spite of the fact that another study has independently found the same reversal of metabolites in a group of hetero- and homosexual subjects (Evans, unpublished), non-specific variables may be producing the reported correlations. Some possibilities could be stress, general activity, recent sexual activity (extent, not type), etc. The third study compared a small number of female as well as male heterosexuals and homosexuals. Levels of androgen were higher and estrogen lower in four homosexual compared to heterosexual females, while androgen was lower than normal in two homosexual males (Lorraine et al., 1970).

The most compelling study to date has compared 30 young adult male homosexuals with 50 male heterosexuals for plasma testosterone levels and additionally has examined the semen of the homosexuals. Those males who were exclusively or almost exclusively homosexual had testosterone levels approximately one half that of the heterosexuals. Additionally, there was a significant correlation between sperm count and degree of homosexuality, with fewer sperm being associated with a greater degree of homosexual orientation (Kolodny et al., 1971). Again, caution must be observed pending confirmation on other subjects with rigorous attention paid to the control of possibly confounding variables, such as stress. It could be, for example, that greater stress experienced by homosexuals because of societal prohibitions on their behavior, influences the findings. Evidence exists from other studies that stress lowers the secretion rate of testosterone. In the male rodent, for example, exposure to a variety of stressors not only lowers plasma and urine testosterone concentrations but also decreases testicular size (Christian, 1955; Bardin and Peterson, 1967). Additionally, Rose et al. (1969) have found lower testosterone excretion levels in soldiers under conditions of both basic training and actual combat.

The additional dimension required in differentiating the male, described above, may help explain why, at the clinical level, psychosexual anomalies are commoner in males (e.g., homosexuality, fetishism, transsexualism, pedophilia, sadism, voyeurism, etc.). In a dual system in which one path automatically evolves and the alternate requires specific influences at specific intervals, more errors are probably along the latter path. (An additional non-hormonal hurdle for the male child may be the necessity of differentiating himself from the first person with whom he is intimate - a female (Greenson, 1967).

Study of the influence gonadal hormone may exert during anatomic and psychosexual differentiation has been greatly facilitated by the increased precision with which these compounds can now be measured. It was quite recent history that testosterone was not measurable in samples of human blood. A breakthrough occurred when gross measures could be made on samples of relatively large volume. An era is now opening in which accurate determinations can be made from the drop of blood emitting from a pin prick. The science has evolved through milligram levels of sensitivity to the microgram (1/1000 milligram) to the nanogram (1/1000 microgram) and currently to the picogram (1/1000 nanogram).

Understanding the role of gonadal hormones is being rendered even more complex by the growing knowledge of their various metabolic pathways. Specific metabolites appear to act at specific body sites. Thus one androgen may be critical for masculine differentiation of the genital system and another for defeminizing specific areas of the brain (Goldfoot, Feder and Goy, 1969; Luttge and Whalen, 1970, 1971).

Precision assays of hormonal metabolites, coupled with recently introduced obstetrical procedures for sampling amniotic fluid and placenta-derived blood may reveal a first look at the prenatal hormonal milieu in man. It may then become possible to correlate this intrauterine endocrine status with postnatal, longitudinal assessments of psychosexual development.

Finally, a word about atypical patterns of sex chromosomes and their relation to atypical sexual behavior. It is a comparatively recent development that chromosomes have become individually visible. Only within the last few years have the consequences of omissions and excesses of chromosomal elements become known. As the incidence of sex chromosome anomalies in the male is about 1/500, the number of persons so affected becomes quite considerable. Controversy exists over the possibly causal interrelation of these chromosomal abnormalities and cross-gender behavior. Several patients have been described with an extra "X" chromosome whc are also transvestites or transsexuals (Money and Pollitt, 1964; Baker and Stoller,

1968). However, it is difficult to rule out sampling bias as the
numbers are small, and such patients are more likely to find their
way into the literature. It is also difficult to control for the
influence of the somatic manifestations of Klinefelter's syndrome
(gynecomastia, small genitalia) on a male's self-concept. The
issue may be settled by prospective studies underway in which males
identified at birth as having an extra X chromosome are undergoing
longitudinal psychologic study (Walzer, S., personal communication).

The importance of direct chromosomal observation, notwith-
standing, it may historically come to be but a small beginning. The
person with testicular feminization has a normal male chromosomal
configuration (44 + XY). Yet, hidden within a normal appearing
chromosome is an invisible defect which renders that male incapable
of realizing its masculine potential.

CONCLUSION

The manner in which gonadal hormones, brain anatomy, and sexual
behavior are interrelated defies precise description. If man were
solely dependent on relatively simple chemical-cellular interactions,
responding in a relatively more programmed manner as in lower animals, the problem of delineating the mechanisms operating in the
above conditions would be difficult enough. In the human, overlaid
with the profound influences of a lifetime of interpersonal experiences, mediated by a more sophisticated central nervous system
network, the task of orderly arranging all the operant influences
becomes insurmountable. For the present we must content ourselves
with descriptions of case reports which alert us to the finding that
striking relationships between gonadal hormones, anatomical structures and sexual behavior exist. Each new research finding merely
enlarges the complexity with which the relationship can be viewed.
It would be as equally erroneous to accept a purely neuroanatomic
or neuroendocrine basis of human sexual behavior as it would be to
discard all the above findings as irrelevant to Man and inconse-
quential when viewed against psychodynamic or learning theory form-
ulations.

NOTES

*Review material in this chapter will appear as a chapter in
(tentatively titled) Cross-Sexed Identity in Children and Adults,
copyright 1972, Richard Green, M.D., to be published by Basic Books,
Inc., and Gerald Duckworth and Co., Ltd.

The author's research is supported by NIMH Research Scientist
Development Award K01 MH31-739 and Foundations' Fund for Research
in Psychiatry Grant G69-471.

REFERENCES

Baker, H. and Stoller, R. 1968. Sexual psychopathology in the
  hypogonadal male. Archiv. of Gen. Psychiat., 18, 361-364.
Ball, J. 1968. A case of hair fetishism, transvestism, and organic
  cerebral disorder. Acta Psychiat Scandinav., 44, 249-254.
Bardin, C. and Peterson, R. 1967. Studies of androgen production
  by the rat. Endocrinol., 80, 38-45.
Blumer, D. 1969. Transsexualism, sexual dysfunction and temporal
  lobe disorder. In: Transsexualism and Sex Reassignment, R. Green
  and J. Money (Eds.), The Johns Hopkins Press, Baltimore.
Christian, J. 1955. Effect of population size on the adrenal
  glands and reproductive organs of male mice in populations of
  fixed size. Amer. J. Physiol., 182, 292-301.
Ehrhardt, A., Epstein, R. and Money, J. 1968. Fetal androgens and
  female gender identity in the early-treated adrenogenital syndrome.
  Johns Hopkins Med. J., 122, 160-167.
Erhardt, A., Evers, K. and Money, J. 1968. Influence of androgen
  and some aspects of sexually dimorphic behavior in women with the
  late-treated adrenogenital syndrome. Johns Hopkins Med. J., 123,
  115-122.
Ehrhardt, A. and Money, J. 1967. Progestin-induced hermaphroditism:
  IQ and psychosexual identity in a study of ten girls. J. Sex
  Research, 3, 83-100.
Freemon, F. and Nevis, A. 1969. Temporal lobe sexual seizures.
  Neurology, 19, 87-90.
Freud, S. 1920. The psychogenesis of a case of homosexuality in
  a woman. In: The Standard Edition of the Complete Psychological
  Works of Sigmund Freud, J. Strachey (Ed.), Hogarth Press, London,
  1955, p. 171.
Gillespie, A. 1971. Paper read at the Second International
  Congress on Gender Identity, Elsinore, Denmark.
Goldfoot, D., Feder, H. and Goy, R. 1969. Development of bisex-
  uality in the male rat treated neonatally with androstenedione.
  J. Comp. Physiol. Psychol., 67, 41-45.
Grady, K., Phoenix, C. and Young, W. 1965. Role of the developing
  testis in differentiation of the neural tissues mediating mating
  behavior. J. Comp. Physiol. Psychol., 59, 176-182.
Greenson, R. 1967. Dis-identifying from mother: Its special imp-
  ortance for the boy. International Psycho-Analytical Congress.
Harris, G. 1964. Sex hormones, brain development, and brain function. Endocrinology, 75, 627-648.
Jones, J. 1971. Paper read at the Second International Congress
  on Gender Identity, Elsinore, Denmark.
Jost, A. 1947. Recherches sur la differenciation sexuelle de l'
  embryo de la lapin. Arch. Anat. Microscop. et Morphol. Exper.,
  36, 151-200; 242-270; 271-315.
Kolarsky, A., Freund, K., Machek, J. and Polak, G. 1967. Male
  sexual deviation. Arch Gen. Psychiat., 17, 735-743.

Kluver, H. and Bucy, P. 1939. Preliminary analysis of functions of the temporal lobes of monkeys. Arch. Neurol. and Psychiat., 42, 979-1000.

Kolodny, R., Masters, W., Hendryx, J. and Toro, G. 1971. Plasma testosterone and semen analysis in male homosexuals. New Eng. J. Med., 285, 1170-1174.

Lorraine, J., Ismail, A., Adamopoulos, A. and Dove, G. 1970. Endocrine function in male and female homosexuals. Brit. Med. J., 4, 406-408.

Luttge, W. and Whalen, R. 1970. Dihydrotesterone, androstenedione, testosterone: Comparative effectiveness in masculinizing and defeminizing reproductive systems in male and female rats. Hormones and Behavior, 1, 265-281.

Migeon, C., Rivarola, M. and Forest, M. 1969. Studies of androgens in male transsexual subjects. In: Transsexualism and Sex Reassignment, R. Green and J. Money (Eds.), The Johns Hopkins Press, Baltimore.

Mitchell, W., Falconer, M. and Hill, D. 1954. Epilepsy with fetishism relieved by temporal lobectomy. Lancet, 2, 626-630.

Margolese, S. 1970. Homosexuality: A new endocrine correlate. Hormones and Behavior, 1, 151-155.

Money, J. 1968. Sex Errors of the Body, The Johns Hopkins Press, Baltimore.

Money, J., Ehrhardt, A. and Masica, D. 1968. Fetal feminization induced by androgen insensitivity in the testicular feminizing syndrome. Johns Hopkins Med. Bull., 123, 105-111.

Money, J. and Pollitt, E. 1964. Cytogenetic and psychosexual ambiguity: Klinefelter's Syndrome and transvestism compared. Arch. Gen. Psychiat., 11, 589-595.

Rivarola, M., Saez, J., Meyer, W., Kenney, F. and Migeon, C. 1967. Studies of androgens in the syndrome of male pseudohermaphroditism with testicular feminization. J. Clin. Endocrin. Metab., 27, 371-378.

Roeder, F. and Muller, D. 1969. The stereotaxic treatment of pedophilic homosexuality. German Med, Monthly, 14, 265-271.

Rose, R.M., Bourne, P. and Poe, R. 1969. Androgen responses to stress. Psychosom. Med., 31, 418-436.

Simmer, H., Pion, R. and Dignam, W. 1965. Testicular Feminization. Charles C. Thomas, Springfield.

Terzian, H. and Dalle Ore, G. 1955. Syndrome of Kluver and Bucy reproduced in man by bilateral removal of temporal lobe. Neurology, 5, 373-380.

Walinder, J. 1965. Transvestism, definition and evidence in favor of occasional derivation from cerebral dysfunction. Internat. J. Neuropsychiat., 1, 567-573.

Whalen, R. and Luttge, W. 1971. Testosterone, androstenedione and dihydrotestosterone: Effects on mating behavior of male rats. Hormones and Behavior, 2, 117-125.

Whalen, R., Peck, C. and LoPiccolo, J. 1966. Virilization of

female rats by prenatally administered progestin. Endocrinology,
    78, 965-970.
Young, W., Goy, R. and Phoenix, C. 1964. Hormones and sexual
    behavior. Science, 143, 212-218.

STEROID HORMONES AND THE CHEMISTRY OF BEHAVIOR

Bruce S. McEwen

The Rockefeller University, New York, New York

INTRODUCTION

One of the striking features of the brain in contrast to other
organs of the body is the extensive regional differentiation of its
structure and function. This has been known for many years by neur-
oanatomists, neurophysiologists, and physiological psychologists
who have explored it using such techniques as lesioning, electrical
and chemical stimulation, and electrical recording. The regional
differentiation of the brain is also increasingly apparent to neur-
ochemists, who have discovered regional differences in the concentration of both large and small molecules within the central nervous system. One of the most striking examples of this differentiation is the distribution of the catechol and indole amines which
are produced by neurons with cell bodies localized primarily within
the midbrain, lower brainstem and hypothalamus, and with nerve
endings which spread throughout the entire central nervous system
(Fuxe, Hökfelt, and Ungerstedt, 1970). Another even more recent
example, which is the subject of this chapter, is the regional distribution of proteins which stereospecifically bind steroid hormones
in a manner that strongly suggests that the hormone is influencing
the activity of the genome in the cell nucleus and thereby influencing neural processes underlying behavior.

Steroid hormones are useful tools with which to study both
biochemical and behavioral aspects of brain function. We have chosen them because of extensive information which shows that steroid
hormones directly affect neural processes underlying behavior and
because of biochemical information for non-neural target tissues
which shows that steroid hormones act on genomic activity in these
tissues to modify the composition and function of the tissue. An

extensive review of these two subjects may be found in McEwen,
Zigmond and Gerlach (1972). Briefly stated, the evidence that hormones directly affect brain function comes from four types of investigations. First, implantation of minute quantities of a hormone into particular brain regions has been found to mimic some of
the effects of systemic administration of larger quantities of the
hormone. Examples to be considered in subsequent sections of this
chapter include the facilitation by implanted estrogen of female
sexual behavior and the alteration by implanted corticosterone of
ACTH secretion. Second, recording of single or multiple unit activity in selected brain regions has shown enhanced or depressed bioelectric activity resulting from systemic or local administration
of steroid hormones. Third, studies of uptake of radioactive steroid hormones have shown: 1) that virtually all steroids enter the
brain from the blood; 2) that some hormones concentrate selectively
in particular brain regions due to the existence in these regions
of limited-capacity binding sites. One of the points of this arti-
cle is to show that hormone binding sites exist in brain regions
where the hormone produces a physiological effect. Fourth, biochemical studies show that certain enzymes are regulated by steroid
hormones and that other, less well-understood, metabolic changes
occur as a result of alterations in levels of particular hormones
in the blood.

ESTRADIOL: TAKEN UP WHERE IT ACTS

When implanted directly into the brain of ovariectomized female
animals, estradiol and the synthetic estrogen, stilbestrol, restore
female sexual receptivity (Lisk, 1962; Michael, 1966). Estradiol
implants also suppress the secretion of gonadotrophic hormone (for
review, see McEwen et al., 1972). There is general agreement that
the most effective brain regions for obtaining these effects are
located within the preoptic region and hypothalamus, including the
"hypophysiotrophic area" defined by Halasz et al. (1962, 1965).
Studies of the uptake from the blood and binding of radioactive
estradiol revealed that these same regions of the brain accumulate
more hormone than any other part of the brain (Eisenfeld and Axelrod, 1967; Kato and Villee, 1967; McEwen and Pfaff, 1970). Estradiol accumulation in preoptic region, hypothalamus, amygdala, and
pituitary, but not in cerebral cortex, is reduced by injections of
unlabelled estradiol-17β, but not by estradiol-17α or testosterone
(Eisenfeld and Axelrod, 1965; Kato and Villee, 1967; McEwen and
Pfaff, 1970). This result demonstrates that the concentration of
estradiol in hypothalamus, preoptic region, pituitary, and amygdala
over that in cortex is due to limited-capacity binding sites which
are specific for the naturally-occurring and active estrogenic steroid, estradiol-17β.

The biochemical properties of these binding sites was revealed
by further studies using cell fractionation techniques and various

biochemical procedures. Eisenfeld (1970) reported finding soluble
proteins which bind estradiol-17β in a stereospecific manner and
which are concentrated in the preoptic region, hypothalamus, and
pituitary. Similar findings have been reported by other laboratories (Kahwanango et al., 1969; Notides, 1970; Kato et al., 1970a;
Mowles et al., 1971). The cell nucleus is another important site
of estradiol binding in the hypothalamus, preoptic area, amygdala
(Zigmond and McEwen, 1970) and pituitary (King et al., 1965; Kato
et al., 1970b). Zigmond and McEwen (1970) found that nearly 40%
of the labelled estradiol in the hypothalamus and preoptic region
at two hours after systemic injection of the labelled hormone is
associated with cell nuclei. Moreover, the nuclear binding sites
have a limited capacity for the hormone and are stereospecific for
estradiol-17β as opposed to the stereoisomer estradiol-17α (a very
weak estrogen) and testosterone (Zigmond and McEwen, 1970). The
potent synthetic estrogen, stilbestrol, which is chemically quite
different from estradiol, does bind to brain nuclear binding sites
(Zigmond, 1971). Nuclear binding of radioactive estradiol has been
reported by other laboratories (Chader and Villee, 1970; Mowles et
al., 1971). The relationship between the soluble and nuclear binding sites for estradiol is not clear, but it is tempting to suppose,
by analogy with the two-step uptake mechanism proposed for uterus
(Jensen et al., 1969), that soluble binding factors for estradiol
carry the hormone to the nuclear binding sites. Some recent evidence concerning the time course of the soluble and nuclear binding
is consistent with this model (Mowles et al., 1971).

     Concurrently with the initial biochemical studies of estradiol
uptake, autoradiographic investigations were also providing evidence
that estradiol accumulates in brain regions where it acts. These
studies demonstrated that more estradiol accumulates in neurons
than in neuropil and that neurons in the preoptic region and hypothalamus accumulate more radioactive material than neurons in other
brain regions (Attramadal, 1965; Michael, 1965; Pfaff, 1968; Stumpf,
1968; Anderson and Greenwald, 1969). Refinement of the autoradiographic technique to minimize diffusion of the labelled hormone
within the cell confirmed the localization of much estradiol in the
cell nuclei (Stumpf, 1968; Anderson and Greenwald, 1969; McEwen et
al., 1972). It can be estimated from the biochemical data concerning estradiol binding that there are, at saturation, 1200 molecules
of estradiol bound per nucleus in the hypothalamus and preoptic
region (McEwen et al., 1972). In view of the autoradiography, showing that not every cell nucleus within these regions binds estradiol,
it is interesting to estimate the number of molecules of estradiol
bound in those cell nuclei which do bind the hormone. If one of
every ten nuclei bound the hormone, there would be 12,000 molecules
per nucleus. Such a figure is of the same order of magnitude as
figures quoted for the number of estrogen binding sites in uterine
(16,000) or anterior pituitary (12,000) cells (Notides, 1970).

Not all regions of the brain where estradiol binds and acts are located within the preoptic area or hypothalamus. For example, neurons in the amygdala accumulate radioactive estradiol, some of which binds to cell nuclei (McEwen and Zigmond, 1970; Stumpf, 1970). The amygdala has been implicated as an area of estrogen sensitivity. Tindal et al. (1969) evoked a lactogenic response after implanting estradiol into the amygdala and stria terminalis but not after implanting it into the pyriform cortex or other brain regions, suggesting that estrogen-sensitive neurons in the amygdala may participate in the regulation of pituitary prolactin secretion. In addition to the amygdala, it is quite likely that other areas of the brain contain neurons which bind and respond to estradiol. This is apparent from the autoradiographic work of Pfaff (1968) and from the biochemical studies of Eisenfeld (1970). McEwen and Pfaff (1970), and Zigmond and McEwen (1970) in which lesser but still significant levels of limited-capacity estradiol uptake and binding were observed outside of the hypothalamus and preoptic area in structures such as the hippocampus and cerebral cortex. In considering possible action of estradiol in these and other brain regions, it should be borne in mind that this hormone produces other neural and behavioral effects besides those so far considered. These effects include the suppression of food intake (Valenstein et al., 1967; Beatty et al., 1970; Wade and Zucker, 1970) and the priming of the LH surge in ovulation (Labhsetwar, 1970; Raziano et al., 1971).

C. CORTICOSTERONE: TAKEN UP DIFFERENTLY FROM ESTRADIOL

The success in correlating estradiol uptake sites with sites of action of this hormone led us to investigate another hormone, corticosterone, which is the principal secretion of the adrenal in the rat. Corticosterone is known to act in a number of brain regions to alter the secretion of ACTH either in response to stress or in connection with the daily rhythm of rest and activity (See Section E). However, the precise sites of action of corticosterone are less clearly known; thus the studies of corticosterone uptake were undertaken as a means of revealing brain structures which accumulate the hormone so that they might be subjected to more careful physiological examination.

As in the case of estradiol uptake, the first studies were conducted with tissue samples from various brain regions (McEwen et al., 1969). The accumulation of radioactivity injected as corticosterone was fairly uniform among brain regions when untreated rats were used; but, after bilateral adrenalectomy, ^{3}H-corticosterone injected into the systemic circulation was found to accumulate in the hippocampus at two to four times higher concentrations than in other brain regions. This enhanced hippocampal uptake after adrenalectomy could be suppressed by injecting with the labelled hormone unlabelled corticosteroids such as corticosterone, hydrocortisone, or dexamethasone, leading us to suppose that corticosterone secretion by the adrenals in untreated animals is normally saturating the uptake sites in brain regions such as hippocampus (McEwen et al., 1969).

The subcellular fractionation of tissue from hippocampus and other brain regions produced further evidence for the existence of corticosterone binding sites in rat brain. As is the case for estradiol, ^{3}H-corticosterone injected systemically 30 minutes to 2 hours before sacrifice was found to be tightly bound to cell nuclei isolated from the brain regions which showed the largest whole tissue accumulation of this hormone (McEwen et al., 1970). The hippocampus had the highest uptake, followed by the amygdala and the cerebral cortex. The hypothalamus, midbrain and brainstem, and cerebellum were all much lower; nevertheless, the binding in these regions was also saturable by unlabelled corticosterone administered with the labelled hormone. In addition to cell nuclear binding of corticosterone, soluble macromolecules were also found which bind corticosterone. These macromolecules are distributed in much the same way as the nuclear binding sites, with highest levels in the hippocampus and lesser amounts in other brain regions. McEwen et al. (1972) have estimated that on the average there are 22,000 corticosterone binding sites in the cell nucleus of hippocampus and around twice that number of soluble binding macromolecules.

Autoradiography of ^{3}H-corticosterone uptake in brains of adrenalectomized rats has supplied additional information which both confirms and extends the biochemical studies of hormone binding (Gerlach and McEwen, 1972). This technique confirms that the hippocampus is the structure with highest uptake, and the label can be seen to concentrate in the region of the cell nucleus in pyramidal neurons of Ammon's horn. Perhaps the most intriguing aspect of the labelling of the hippocampus is that some neurons are labelled very heavily, while others are not appreciably labelled at all. The heavily labelled neurons are in clusters in CA1 and CA2 of Ammon's horn; the largest cluster occurring in CA2, CA3 and CA4 are much less heavily labelled. In the dentate gyrus the radioactive material is concentrated in certain neurons which are scattered throughout this layer among numerous lightly labelled cells. These observations suggest that there has been a biochemical differentiation of neurons within the hippocampus with respect to corticosterone binding ability. We must await further anatomical and physiological information to know whether this differentiation has a functional significance. One piece of evidence which supports a functional differentation is the finding of Kawakami et al. (1968) that implantations of corticosterone into rabbit hippocampus showed CA2 to be the most effective site for increasing basal ACTH secretion.

D. PHYSICAL AND CHEMICAL PROPERTIES OF HORMONE BINDING MATERIAL IN
BRAIN

Determination of the physical and chemical properties of the
hormone binding materials in brain is essential so that we can
answer a number of important questions about hormone binding and
action in the brain. The first question concerns the mechanism of
hormone uptake and binding to the cell nucleus and the specificity
of this process to a particular tissue. Study of estrogen binding
to brain, pituitary, and uterus offers some hope of answering these
questions. Soluble estrogen–binding macromolecules in hypothalamus
have been characterized as proteins with sulfhydryl groups which are
essential for hormone binding (Eisenfeld, 1970). In this respect
they resembel the soluble estrogen–binding proteins from uterus
(Jensen et al., 1967). Also with respect to their sedimentation
properties in the ultracentrifuge, soluble estrogen–binding proteins
from hypothalamus, pituitary, and uterus appear to be similar. Soluble binding proteins in the uterus have been reported to have "S"
values of 8 to 9 (Toft and Gorski, 1966; Jensen et al., 1967).
Pituitary cytosol proteins have also been reported to have "S" values
of 8 to 9 (Kahwanango et al., 1969; Kato et al., 1970a; Notides,
1970), although one laboratory has reported a value of 4.5 S (Mowles
et al., 1971). Hypothalamic soluble estrogen–binding proteins have
been described as sedimenting with the same "S" value as those from
pituitary (Kahwanango et al., 1969). Estrogen–binding proteins can
also be extracted from cell nuclei with 0.4 M salt solutions and
the sedimentation properties of these molecules have been studied
in the ultracentrifuge. Nuclear extracts of the uterus give "S"
values of 5 (Jensen et al., 1967; Puca and Bresciani, 1968). Pituitary and hypothalamic nuclear extracts give "S" values of 6 to 7
(Kato et al., 1970a; Mowles et al., 1971), which would seem to be
different from the uterus. If indeed this is the case and if further
experiments show differences in specificity of binding to nuclei
among estrogen–binding proteins from uterus, pituitary, and brain,
we shall be able to understand the specificity of action of estrogen
on each of these three tissues in terms of the specific structure and
sites of attachment of the binding protein within the cell. On the
other hand, if all estrogen–binding proteins, soluble and nuclear,
are shown to consist of basically the same proteins, then we must
look further for the basis of tissue-specific actions of the hormone.

Another biochemical question concerning the binding of hormones
to the brain concerns the relationship between the tissue binding
substances and those in the blood, which carry the hormone to its
target tissue. This problem is particularly well illustrated by the
corticosterone-binding proteins in blood and the brain. The principal blood binding protein, which has been called transcortin or
corticosteroid-binding globulin (CBG), binds corticosterone with
high capacity and reasonably high affinity. Several attempts were

made to distinguish between protein and the soluble corticosterone-
binding protein in the brain (McEwen et al., 1972). First, brains
were perfused at sacrifice with dextran-saline solution in order to
remove all blood. ^{3}H-corticosterone was still found to be bound
tightly to cytosol protein and to isolated cell nuclei, thus ruling
out the possiblity that blood binding of corticosterone was responsible for the soluble binding factor in brain extracts. Secondly,
we found that the brain cytosol-binding protein, labelled _in_ _vivo_
with ^{3}H-corticosterone, could be quantitatively precipitated with
hormone attached by protamine sulfate. Serum-binding material was
not so precipitated, even when mixed with the cytosol material.
Third, we found that the brain cytosol-binding protein could be completely separated from serum-binding material by electrophoresis in
acrylamide gels containing 10% glycerol. In these gel separa-
tions, the serum-binding material migrated much faster than brain
binding protein. Similarly, centrifugation of serum and cytosol in
glycerol density gradients revealed that the brain binding protein
is larger in molecular weight than serum-binding factor, with an
apparent "S" value of "5" compared to "3.5" for serum factor. When
gradients were run in the presence of 0.4 M NaCl, the cytosol protein appeared to an "S" value of "3" (corresponding to half the
molecular weight of a "5 S" protein) and to be lighter (or smaller)
than the serum-binding material. It therefore appears that cytosol-
and serum-binding proteins are distinguishable. Whether in fact
they are totally different proteins is not yet clear.

    Yet another contribution of biochemical studies to our under-
standing of hormone binding by the brain is illustrated by experiments conducted on the exchange of radioactive corticosterone already
bound to the binding protein with unlabelled corticosterone added
to the medium (McEwen et al., 1972b). Three types of binding were
compared: serum, brain cytosol, and brain nuclear. During two
hours of exchange at 4°C, the serum exchanged 95% of its labelled
hormone with unlabelled corticosterone in the medium; the cytosol
exchanged 18%, while the nuclei did not exchange any labelled hormone. During 20 hours at the same temperature, the cytosol exchanged
44% of its radioactive hormone while the nuclei still did not show a
detectable exchange. In so far as the exchange reveals the relative
affinity of the binding site for the hormone, it shows that the
sequence of uptake should proceed naturally in the direction of
blood to cytosol to nuclei. As noted in the previous section, proof
is lacking that the brain cytosol binding actually is essential for
and precedes the brain nuclear uptake.

    It is interesting that in spite of the extreme stability of
corticosterone binding to brain cell nuclei, isolated and maintained
at 4°C, the binding of the hormone _in_ _vivo_ is in a very dynamic
state. This is shown by two experiments. First, the time course of
radioactive corticosterone binding to hippocampal cell nuclei shows

a peak of binding at one hour followed by a rapid fall after two
hours to almost no bound radioactivity four hours after the hormone
injection (McEwen et al., 1970). The binding of estradiol to
hypothalamic-preoptic cell nuclei shows a similar peak at one hour
after hormone injection with a somewhat slower but still pronounced
loss in the next three hours (Zigmond and McEwen, 1970; Mowles et
al., 1971). Another experiment illustrated the turnover of bound
corticosterone even more dramatically.

In this experiment, radioactive corticosterone was injected
into adrenalectomized rats and was followed 20 minutes later by an
injection of 3 milligrams of unlabelled corticosterone. In previous
experiments, this dose of unlabelled hormone given before the labelled hormone would completely block binding of the radioactive hormone
to the binding sites (McEwen et al., 1969, 1970), but might be
expected when injected after the radioactive hormone not to interfere with its binding. We found, however, that it did interfere
and "washed out" the nuclear and cytosol binding which we normally
would see when the animals were killed at one hour after the injection. Several factors must be considered in understanding the
mechanism by which hormone binding to a cellular binding protein is
terminated. The time course and "wash-out" experiments, taken together, suggest that for corticosterone there is a fairly rapid exchange
of hormone between the blood and the binding site. Another factor
in determining the extent or duration of binding may be the metabolism of the hormone within the cell or by adjacent cells. We have
found, for example, that at one hour after injecting radioactive
corticosterone 90% of the bound is unmetabolized, but less than 60%
of the unbound radioactivity is unmetabolized (McEwen et al., 1972b).

E. EVIDENCE FOR A ROLE OF CORTICOSTERONE IN HIPPOCAMPUS

Localization of sites of action of estradiol within the brain
led to studies which demonstrated hormone uptake and binding by
these same brain regions. Stimulated by these studies, investigations of corticosterone uptake and binding by brain regions have
provided us with a phenomenon, namely, the intense concentration of
this hormone by the hippocampus. Now we must ask the question:
What is the predictive value of this observation, i.e., what are
the effects of corticosterone on hippocampal function?

Preliminary answers to this question may be obtained both from
the neuroendocrine literature dealing with the regulation of ACTH
secretion and from the psychological literature pertaining to hippocampal function. Neuroendocrinologists studying the regulation
of ACTH secretion found that lesions in the hippocampus or sectioning of the fornix (main output from hippocampus) tended to abolish
the normal diurnal variation of ACTH secretion (Mason, 1958; Nakadate
and DeGroot, 1963; Moberg et al., 1971). Maintenance of constant

levels of adrenal steroid by implantation of cortisone in hippocampus was found to abolish the diurnal ACTH variation (Slusher, 1966),
thus indicating a role for adrenal steroids in the maintenance of
this diurnal cycle. This was further supported by the observation
that the diurnal cycle of ACTH secretion, though not abolished by
adrenalectomy, shifts to a peak earlier in the day after bilateral
adrenalectomy (Chiefetz et al., 1968; Hiroshige and Sakakura, 1971).
In this connection, bilateral adrenalectomy did abolish the circadian rhythm of paradoxical sleep distribution and injection of
cortisol in late afternoon restablished a diurnal variation in this
parameter (Johnson and Sawyer, 1971).

With respect to the regulation of ACTH secretion at rest and
during stress, various investigators have reported that implantation of corticosteroids in hippocampus enhance both basal and
stress-induced ACTH secretion (Davidson and Feldman, 1967; Knigge,
1967; Bohus et al., 1968; Kawakami et al., 1968b). Since these
effects are opposite to the generally inhibitory effects of direct
hippocampal stimulation on ACTH release (Mason, 1958; Mason et al.,
1962; Okinaka, 1962; Mangili et al., 1966; Kawakami et al., 1968a),
it has been suggested that corticosteroids act on the hippocampus
to inhibit a functionally inhibitory system in this structure which
is concerned with the blocking or shutting of inputs to the hypothalamus which stimulate ACTH release (McEwen et al., 1972).
Other brain structures, such as the septum, amygdala, midbrain, and
hypothalamus, have been implicated in the control of ACTH release,
and direct effects of corticosteroids have been shown by implantation (for reviews see Mangili et al., 1966; McEwen et al., 1972).
Thus the hippocampus is not the only structure governing ACTH
release that is sensitive to adrenal steroid "feedback." It is
interesting that studies of binding of ^{3}H-corticosterone show that
all of these brain regions besides hippocampus, which are implicated
in ACTH release, have some binding sites for this hormone (McEwen
et al., 1970).

The hippocampus is of great importance to acquisition and
successful performance of certain types of behavior, especially
passive avoidance, time delay or spatial alternation of responding,
and partially-reinforced responding ("frustrative non-reward")
(Douglas, 1967; Kimble, 1968; Gray, 1970). During times of hippocampal activity in the rat there often occurs in the hippocampus a
theta rhythm of fixed frequency, 7.5 to 8.5 Hz (Gray, 1970). The
theta rhythm originates in the hippocampus under influence of pace-
maker cells located in the medial septum (Stumpf, 1965). This
rhythm may be an indication of functional activity of the hippocampus (Gray, 1970), although it also has been interpreted as a sign of
inactivity (Grastyan, 1959; Douglas, 1967). Quite possibly, hippocampal activity may be concerned with inhibition of neural activity
associated with behavior in other brain structures (Gray, 1970),

much as the hippocampus plays an inhibitory role in modulating ACTH
secretion.

The functional importance of the septo-hippocampal system and
theta rhythm to behavioral adaptations is suggested by a series of
experiments conducted on the phenomenon known as "frustrative non-
reward" (enhanced responding of partially reinforced subjects compared to continuously reinforced subjects). Drugs such as amobarbital and alcohol prevent the effects of frustrative non-reward
(Miller, 1964; Wagner, 1966; Gray, 1967). Gray and Ball (1969)
provided an attractive mechanism for the action of amobarbital,
which ties in with septal and hippocampal function. They found
that this drug raises the threshold for the appearance of the 7.5
to 8.5 Hz component of theta which occurs during frustrative non-
reward and proposed that the behavioral effects of amobarbital are
due to this disruption. Extension of this line of investigation has
brought the phenomenon closer to the question of adrenal influences
on hippocampal function. Gray et al. (1971) found that ACTH administered to rats during acquisition of partially reinforced appetitive behavior prevented the development of the typical increased
response rate ("frustrative non-reward"). Not only this, but such
ACTH treatments during acquisition appear to block effects of subsequent ACTH treatment during extinction in retarding the rate of
extinction (Gray, 1971).* These effects of ACTH are opposite to
the effects of electrically driving the theta rhythm in the septum
during acquisition and extinction and suggest that this hormone is
directly or indirectly acting like amobarbital (Gray et al., 1971).
it is not yet known whether these ACTH effects are direct or are
mediated by the secretion of adrenal steroids.

Another aspect of the hippocampus in relation to behavior is
the sensitivity of this structure to various agents capable of producing retrograde amnesia, such as direct electrical stimulation
(Nyakas and Endroczi, 1970; Pagano et al., 1970), spreading depression (Avis and Carlton, 1968; Auerbach and Carlton, 1971), and local
administration of protein synthesis inhibitors (Flexner et al.,
1967; Daniels, 1971). Electrical stimulation of the hippocampus
has been shown to produce effects on both behavioral and neuroendocrine parameters: not only was avoidance behavior shown to be absent
following such stimulation, but also ACTH secretion associated with
the conditioned stimulus was missing (Nyakas and Endroczi, 1970;
Auerbach and Carlton, 1971). The basis of this amnesic effect appears to be propagated, seizure-like discharges from the hippocampus
into the midbrain region (Nyakas and Endroczi, 1970). Other amnesic
agents such as puromycin have been found to produce seizure-like
activity in hippocampus (Cohen and Barondes, 1967). Interestingly
enough, the ability of this drug to induce amnesia depends on the
presence of the adrenal glands (Flexner and Flexner, 1970). This
finding may be related to the observation that corticosteroids

facilitate the development of stimulation-induced seizure activity
in the hippocampus (Endroczi and Lissak, 1962).

There are obviously many phenomena associated with the hippocampus which have been shown to be influenced by adrenal corticosteroids. We hope eventually to provide the connection between these
phenomena and corticosterone binding in hippocampal neurons. One
piece of evidence which offers a beginning in making this connection
is the observation by Pfaff et al. (1971) that corticosterone administered subcutaneously to rats suppresses the unit activity of neurons in the hippocampus. The effect appears with a delay of 30 minutes to one hour after hormone injection and lasts for at least 3
hours. Both of these temporal characteristics suggest that some
long-lasting metabolic response is involved rather than an immediate
and direct effect of enhanced blood levels of the hormone. Moreover,
the direction of the effect, a reduction of activity, is consistent
with the effects of corticosteroids on hippocampal function influencing ACTH release (see above), namely, to antagonize the activity
of a system in hippocampus which may well actively inhibit neural
activity in the hypothalamus. In these same experiments, ACTH was
found to increase unit activity of many of these same neurons. This
result illustrated at the level of bioelectric activity the opposite
effects of ACTH and adrenal glucocorticoids which have been noted in
behavioral experiments (de Wied, 1967; Bohus, 1970; Weiss et al.,
1970).

F. BIOCHEMICAL EFFECTS OF STEROID HORMONES ON BRAIN REGIONS

We have tried to connect the binding of steroid hormones to
specific neuronal cell nuclear binding sites--potential "receptors"
--with the action of the hormones in these same brain regions by
suggesting that the bound hormone in some way alters genomic activity leading ultimately to the production of proteins which alter the
level or nature of neuronal function (McEwen et al., 1972). These
proteins might be enzymes concerned, for example, with neurotransmitter biosynthesis or breakdown, or with oxidative metabolism underlying the production of high energy reserves, or with ionic balance
in neurons which is so important for proper bioelectric activity.
Specific examples of hormone-induced changes of these kinds of enzymes and their matabolism have already been reviewed in two previous
publications (McEwen et al., 1970; McEwen et al., 1972).

When we find such changes, what can be done to demonstrate their
participation in hormone-stimulated behavior and in neuroendocrine
physiology? Demonstration of genomic involvement in steroid hormone
effects on the brain can be made by the systemic or local injection
of inhibitors of RNA and protein synthesis at the time of administration of the hormone. Some success has been obtained by Schalley
et al. (1969) who were able to block with actinomycin D the effects

of estrogen in suppressing pituitary secretion of gonadotrophic
hormones. It is also possible to interfere with estrogen induction
of female sexual behavior by local implantation in the preoptic area
of RNA and protein synthesis inhibitors (Quadagno et al., 1971).
Moving several steps away from the genome to enzymes and the substrates and products of their action, there are similar pharmacological techniques of administering antagonists or analogs of particular naturally-occurring substances, such as neurotransmitters,
which can be used to demonstrate their importance for hormonally-
regulated neural processes. In neuroendocrinology, studies of this
kind have shown that dopamine stimulates release of the releasing
hormone for pituitary LH secretion and that other amines such as
norepinephrine and serotonin may also play a role in the control
(Schneider and McCann, 1970). Other studies have shown that serotonin (Scapagnini et al., 1971), norepinephrine (Bhattacharya and
Marks, 1969; Van Loon et al., 1971) and acetylcholine (Hedge and
Smelik, 1968) are involved in the regulation of ACTH secretion.
There is no reason why this same pharmacological approach will not
be effective in studies of behavior affected by hormones, since
chemical stimulation of feeding and drinking behavior has been
demonstrated (Miller, 1965). As this kind of experimentation reveals
more and more about the regional pharmacology concerned with neuroendocrine and behavioral regulation, we envision a situation where
we may combine the localization of these effects with the maps showing where steroid hormone binding sites occur in the brain.

G. CONCLUSIONS

The biological phenomena encountered in the study of the binding
and effects of steroid hormones on the brain are relevant to the
three key words of this symposium: mood, motivation, and memory.
Pathological aberrations of mood have been linked to the absence or
excess of adrenal steroids (Coppen, 1967; Frawley, 1967; Liddle,
1967); the normal motivation for sexual contact is strongly dependent upon sex hormones (see Section B); and deficits of memory produced by inhibitors of protein synthesis and by electroconvulsive
shock can be duplicated by local administration of these agents or
local brain stimulation in hippocampus, a brain region where adrenal
steroids concentrate, bind to cell nuclei, and influence neural processes. And, as indicated in Section E, involvement of the pituitary-
adrenal axis in the development of amnesias has been reported.

The study of hormone binding by brain began with the observation that estradiol binds in the regions of the brain where it acts
to affect neural processes underlying neuroendocrine and behavioral
responses related to reproduction. This kind of investigation has
led us to the discovery that another hormone, corticosterone, binds
to different regions of the brain from estradiol. The strong binding
of this hormone in hippocampus suggests that it acts on neural

processes in that structure; and evidence from neuroendocrinology,
physiological psychology, and neurophysiology indicates that this
is the case and that the neural processes affected by corticosterone are related to the neuroendocrine and behavioral responses to
novel or noxious stimuli and to stimuli which cause the animal to
be alerted. The rapid coalescence of information from neurochemistry, physiology, and behavior which is illustrated by these
studies and is representative of much current research in neurobiology indicates that we have entered a new era in our ability to
probe and understand brain function.

*FOOTNOTE

The effects of ACTH which concern acquisition or extinction of
partially reinforced behavior are undoubtedly related to effects of
ACTH observed by Miller and Ogawa (1962) and by de Wied, Bohus and
coworkers (Bohus and de Wied, 1967; De Wied, 1967; Bohus et al.,
1968; De Wied, 1970). These investigators observed that ACTH retards extinction of conditioned avoidance behavior when administered during the extinction period. It is worth noting that the
effects of ACTH and of adrenal steroids can be separated in these
studies: while ACTH retards extinction, adrenal steroids enhance
the rate of extinction (de Wied, 1967; Bohus, 1970).

REFERENCES

Anderson, C.H. and Greenwald, S.S. 1969. Autoradiographic analysis of estradiol uptake in the brain and pituitary of the female
rat. Endocrinology, 85, 1160-1165.
Attramadal, A. 1965. Distribution and site of action of oestradiol
in the brain and pituitary gland of the rat following intramus-
cular administration. In: Proceedings Second Internat. Congress
of Endocrinology, Part 1, pp. 612-616. Excerpta Medica Foundation,
Series No. 83, London.
Auerbach, P. and Carlton, P.L. 1971. Retention deficit correlated
with a deficit in the corticocoid response to stress. Science,
173, 1148-1149.
Avis, H.H. and Carlton, P.L. 1968. Retrograde amnesia produced by
hippocampal spreading depression. Science, 161, 73-75.
Beatty, W.W., Powley, T.L. and Keesey, R.E. 1970. Effects of
neonatal testosterone injection and hormone replacement in adulthood on body weight and body fat in female rats. Physiol. Behav.,
5, 1093-1098.
Bhattacharya, A.N. and Marks, B.H. 1969. Reserpine- and chlorpro-
mazine-induced changes in hypothalamo-hypophyseal-adrenal system
in rats in the presence and absence of hypothermia. J. Pharmacol.
Exp. Therap., 165, 108-116.
Bohus, B. 1970. Central nervous structures and the effect of ACTH
and corticosteroids on avoidance behavior: A study with intra-

cerebral implantation of corticosteroids in the rat. Progress in
    Brain Research, 32, 171–183.
Bohus, B. and de Wied, D. 1967. Avoidance and escape behavior
    following medial thalamic lesions in rats. J. Comp. Physiol.
    Psychol., 64, 26–30.
Bohus, B., Nyakas, C., and Endroczi, E. 1968. Effects of adreno-
    corticotrophic hormone on avoidance behavior of intact and adre-
    nalectomized rats. Int. J. Neuropharmacol., 7, 307–314.
Bohus, B., Nyakas, C. and Lissak, K. 1968. Involvement of supra-
    hypothalamic structures in the hormonal feedback action of corti-
    costeroids. Acta Physiol Hungr., 34, 1–8.
Chader, G. J. and Villee, C.A. 1970. Uptake of oestradiol by the
    rabbit hypothalamus. Biochem. J., 118, 93–97.
Cheifetz, P., Gaffud, N., and Dingman, J.F. 1968. Effects of bi-
    lateral adrenalectomy and continuous light on the circadian
    rhythm of corticotrophin in female rats. Endocrinology, 82, 1117–
    1124.
Cohen, H.D. and Barondes, S.H. (1967) Puromycin effect on memory
    may be due to occult seizures. Science, 157, 333–334.
Coppen, A. 1967. The biochemistry of affective disorders. J.
    Psychiat., 113, 1237–1264.
Daniels, D. 1971. Acquisition, storage, and recall of memory for
    brightness discrimination by rats following intracerebral infu-
    sion of acetoxycycloheximide. J. Comp. Physiol. Psychol., 76,
    110–118.
Davidson, J.M. and Feldman, S. 1967. Effects of extrahypothalamic
    dexamethasone implants on the pituitary adrenal system. Acta
    Endocrinol., 55, 240–246.
De Wied, D. 1967. Opposite effects of ACTH and glucocorticoster-
    oids on extinction of conditioned avoidance behavior. In: Proc.
    Second Internat. Congress on Hormonal Steroids, Milan, pp. 945–
    951. Excerpta Medica Internat. Congress Series, No. 132.
De Wied, D., Witter, A. and Lande, S. 1970. Anterior pituitary
    peptides and avoidance acquisition of hypophysectomized rats.
    Progress in Brain Research, 32, 213–233.
Douglas, R.J. 1967. The hippocampus and behavior. Psychol. Bull.,
    67, 416–442.
Eisenfeld, A.J. 1967. Computer analysis of the distribution of
    [^{3}H] estradiol. Biochim. Biophys. Acta, 136, 498–507.
Eisenfeld, A.J. 1970. ^{3}H-Estradiol: In vitro binding to macro-
    molecules from the rat hypothalamus, anterior pituitary, and
    uterus. Endocrinology, 86, 1313–1318.
Eisenfeld, A.J. and Axelrod, J. 1965. Selectivity of estrogen dis-
    tribution in tissues. J. Pharmacol. Exp. Therap., 150, 469–475.
Endroczi, E. and Lissak, K. 1962. Interrelations between paleo-
    cortical activity and pituitary adrenocortical function. Acta
    Physiol. Hungr., 21, 257–263.
Flexner, J.B. and Flexner, L.B. 1970. Adrenalectomy and the sup-
    pression of memory by puromycin. Proc. Nat. Acad. Sci., 66, 48–52.

Flexner, L.B., Flexner, J.B. and Roberts, R.B. 1967. Memory in
  mice analyzed with antibiotics. Science, 155, 1377-1383.
Frawley, T.F. 1967. Adrenal cortical insufficiency. In: The
  Adrenal Cortex. A.B. Eisenstein (Ed.). Boston: Little, Brown,
  pp. 439-521.
Fuxe, K., Hokfelt, T. and Ungerstedt, U. (1970) Morphological and
  functional aspects of central monoamine neurons. Int. Rev. Neuro-
  biol., 13, 93-126.
Gerlach, J. and McEwen, B.S. 1972. Rat brain binds adrenal steroid
  hormones: Radioautography of hippocampus with corticosterone.
  Science, in press.
Grastyan, E. 1959. The hippocampus and higher nervous activity.
  In: The Central Nervous System and Behavior (Second Conference).
  M.A.B. Brazier (Editor). Josiah Macy, Jr. Foundation, New York.
Gray, J.A. 1967. Disappointment and drugs in the rat. Advancement
  of Science, 23, 595-605.
Gray, J.A. 1970. Sodium amobarbital, the hippocampal theta rhythm,
  and the partial reinforcement extinction effect. Psychol. Rev.,
  77, 465-480.
Gray, J.A. 1971. Effect of ACTH on extinction of rewarded behavior
  is blocked by previous administration of ACTH. Nature (London),
  229, 52-54.
Gray, J.A. and Ball, G.G. 1970. Frequency-specific relation between
  hippocampal theta rhythm, behavior, and amobarbital action.
  Science, 168, 1246-1248.
Gray, J.A., Mayes, A.R. and Wilson, M. 1971. A barbiturate-like
  effect of adrenocorticotrophic hormone on the partial reinforcement acquisition and extinction effects. Neuropharmacology, 10,
  223-230.
Halasz, B., Pupp, L. and Uhlarik, S. 1962. Hypophysiotrophic area
  in the hypothalamus. J. Endocrinol., 25, 147-154.
Halasz, B., Pupp, L., Uhlarik, S. and Tima, L. 1965. Further studies on the hormone secretion of the anterior pituitary transplant-
  ed into the hypophysiotrophic area of the rat hypothalamus.
  Endocrinology, 77, 343-355.
Hedge, G.A. and Smelik, P.G. 1968. Corticotropin release: Inhibition by intrahypothalamic implantation of atropine. Science, 159,
  891-892.
Hiroshige, T. and Sakakura, M. 1971. Circadian rhythm of cortico-
  trophin-releasing activity in the hypothalamus of normal and
  adrenalectomized rats. Neuroendocrinology, 7, 25-36.
Jensen, E.V., Hurst, D.J., DeSombre, E.R. and Jungblut, P.W. 1967.
  Sulfhydryl groups and estradiol-receptor interaction. Science,
  158, 385-387.
Jensen, E.V., Suzuki, T., Numata, M., Smith, S. and DeSombre, E.R.
  1969. Estrogen-binding substances of target tissues. Steroids,
  13, 417-427.
Kahwanango, I., Heinrichs, W.L. and Herrman, W.L. 1969. Isolation
  of estradiol "receptors" from bovine hypothalamus and anterior
  pituitary gland. Nature (London), 223, 313-314.

Kato, J., Atsumi, Y., and Inaba, M.   1970a.   A soluble receptor
    for estradiol in rat anterior hypophysis.   J. Biochem. (Tokyo),
    68, 759-761.
Kato, J., Atsumi, Y., and Muramatsu, M.   1970b.   Nuclear estradiol
    receptor in rat anterior hypophysis.   J. Biochem. (Tokyo), 67,
    871-872.
Kato, J. and Villee, C.A.   1967.   Preferential uptake of estradiol
    by the anterior hypothalamus of the rat.   Endocrinology, 80,
    567-575.
Kawakami, M., Seto, K., Terasawa, E., Yoshida, E., Miyamoto, T.,
    Sekiguchi, M. and Hattori, Y.   1968a.   Influence of electrical
    stimulation and lesion in limbic structure upon biosynthesis of
    adrenocorticoid in the rabbit.   Neuroendocrinology, 3, 337-348.
Kawakami, M., Seto, K. and Yoshida, K.   1968b.   Influence of corti-
    costerone implantation in limbic structures upon biosynthesis of
    adrenocortical steroid.   Neuroendocrinology, 3, 349-354.
Kimble, D.P.   1968.   Hippocampus and internal inhibition.   Psychol.
    Bull., 70, 285-295.
King, R.J.B., Gordon, J. and Inman, D.R.   1965.   The intracellular
    localization of oestrogen in rat tissues.   J. Endocrinol., 32,
    9-15.
Knigge, K.M.   1966.   Feedback mechanisms in neural control of ade-
    nohypophyseal function: Effect of steroids implanted in amygdala
    adn hippocampus.   Abstracts, 2nd Internat. Congr. Hormonal
    Steroids, Milan, p. 208.
Labhsetwar, A.P.   1970.   The role of oestrogens in spontaneous
    ovulation: Evidence for positive oestrogen feedback in the 4-day
    oestrus cycle.   J. Endocrinol., 47, 481-493.
Liao, S. and Fang, S.   1969.   Receptor proteins for androgens and
    the mode of action of androgens on gene transcription in ventral
    prostate.   Vitamins and Hormones, 27, 17-90.
Liddle, G.W.   1967.   Cushing's syndrome.   In: The Adrenal Cortex.
    A.B. Eisenstein, Ed., Little, Brown, Boston, pp. 523-551.
Lisk, R.D.   1962.   Diencephalic placement of estradiol and sexual
    receptivity in the female rat.   Amer. J. Physiol., 203, 493-496.
McEwen, B.S. and Pfaff, D.W.   1970.   Factors influencing sex hor-
    mone uptake by rat brain regions. I. Effects of neonatal treat-
    ment, hypophysectomy, and competing steroid on estradiol uptake.
    Brain res., 21, 1-16.
McEwen, B.S., Magnus, C. and Wallach, G.   1972a.   Soluble cortico-
    sterone-binding macromolecules extracted from rat brain.   Endo-
    crinology, 90, 217-226.
McEwen, B.S., Weiss, J.M. and Schwartz, L.S.   1969.   Uptake of
    corticosterone by rat brain and its concentration by certain
    limbic structures.   Brain Res., 16, 227-241.
McEwen, B.S., Weiss, J.M. and Schwartz, L.S.   1970a.   Retention of
    corticosterone by cell nuclei from brain regions of adrenalecto-
    mized rats.   Brain Res., 17, 471-482.
McEwen, B.S., Zigmond, R.E., Azmitia, E.C., Jr. and Weiss, J.M.

1970b. Steroid hormone interaction with specific brain regions.
   In: Biochemistry of Brain and Behavior. R.E. Bowman and S.P.
   Datta (Eds.). Plenum Press, New York, pp. 123-167.
McEwen, B.S., Zigmond, R.E. and Gerlach, J. 1972b. Sites of ster-
   oid binding and action in the brain. In: Structure and Function
   of the Nervous System, Vol. 4, G.H. Bourne (Ed.), Academic Press,
   New York, in press.
Mangili, C., Motta, M. and Martini, L. 1966. Control of adrenocor-
   ticotrophic hormone secretion. In: Neuroendocrinology, Vol. 1
   L. Martini and W.F. Ganong (Eds.), pp. 297-370.
Mason, J.W. 1958. In: Reticular Formation of the Brain, H.H.
   Jasper et al. (Eds.), Little, Brown, Boston, pp. 645-662.
Mason, J.W., Nauta, W.J.H., Brady, J.V., Robinson, J.A. and Sachar,
   E.J. 1962. The role of limbic system structures in the regula-
   tion of ACTH release. Acta Neurovegitativa, 23, 4-14.
Michael, R.P. 1965. Oestrogens in the central nervous system.
   Brit. Med., Bull., 21, 87-90.
Michael, R.P. 1966. In: The Brain and Gonadal Function, R.A.
   Gorski and R.E. Whalen (Eds.), Brain and Behavior, Vol. 3. Uni-
   versity of California Press, Los Angeles, pp. 82-98.
Miller, N.E. 1964. The analysis of motivational effects illus-
   trated by experiments on amylobarbitone. In: Animal Behaviour
   and Drug Action, H. Steinberg (Ed.), Churchill, London, pp. 1-18.
Miller, N.E. 1965. Chemical coding of behavior in the brain.
   Science, 148, 328-339.
Miller, R.E. and Ogawa, N. 1962. The effect of adrenocortico-
   trophic hormone (ACTH) on avoidance conditioning in the adrenal-
   ectomized rat. J. Comp. Physiol. Psychol., 55, 211-213.
Moberg, G.P., Scapagnini, U., DeGroot, J. and Ganong, W.F. 1971.
   Effect of sectioning the fornix on diurnal fluctuation in plasma
   corticosterone levels in the rat. Neuroendocrinology, 7, 11-15.
Mowles, T.F., Ashkanazy, B., Mix, E., Jr. and Sheppard, H. 1971.
   Hypothalamic and hypophyseal estradiol-binding complexes.
   Endocrinology, 89, 484-491.
Nakadate, G.M. and DeGroot, J. 1963. Fornix transection and adre-
   nocortical function in rats. Anat. Rec., 145, 338.
Notides, A.C. 1970. Binding affinity and specificity of the
   estrogen receptor of the rat uterus and anterior pituitary.
   Endocrinology, 87, 987-992.
Nyakas, C. and Endroczi, E. 1970. Effect of hippocampal stimula-
   tion on the establishment of conditioned fear response in the
   rat. Acta Physiol Hungr., 37, 281-289.
Okinaka, S. 1962. Die Regulation der Hypophysen-Nebennierenfunk-
   tion durch das Limbic-System und der Mittelnhirnanteil der
   Formatio-Reticularis. Acta Neurovegitativa, 23, 15-20.
Pagano, R.R., White, R. and Waters, R.S. 1970. Paper presented
   at Psychonomic Science Meeting, San Antonio, Texas.
Pfaff, D.W. 1968. Uptake of estradiol-17β-^{3}H in the female rat
   brain: An autoradiographic study. Endocrinology, 82, 1149-1155.

Pfaff, D.W., Silva, M.T.A. and Weiss, J.M. 1971. Telemetered recording of hormone effects on hippocampal neurons. Science, 172, 394-395.

Puca, G.A. and Bresciani, F. 1968. Receptor molecule for oestrogens from rat uterus. Nature (London), 218, 967-969.

Quadagno, D.M., Shryne, J. and Gorski, R.A. 1971. The inhibition of steroid-induced sexual behavior by intrahypothalamic actinomycin D. Hormones and Behavior, 2, 1-10.

Raziano, J., Cowchock, S., Merin, M. and Van de Wiele, R.L. 1971. Estrogen dependency of monoamine-induced ovulation. Endocrinology, 88, 1516.

Scapagnini, V., Moberg, G.P., Van Loon, G.R., DeGroot, J. and Ganong, W.F. 1971. Relation of brain 5-hydroxytryptamine content to the diurnal variation in plasma corticosterone in the rat. Neuroendocrinology, 7, 90-96.

Schally, A.V., Bowers, C.Y., Carter, W.H., Arimura, A., Redding, T.W. and Saito, M. 1969. Effect of actinomycin D on the inhibitory response of estrogen on LH release. Endocrinology, 85. 290-299.

Schneider, H.P.G. and McCann, S.M. 1970. Mono- and indolamines and control of LH secretion. Endocrinology, 86, 1127-1133.

Slusher, M.A. 1966. Effects of cortisol implants in the brainstem and ventral hippocampus on diurnal corticosterone levels. Exp. Brain. Res., 1, 184-194.

Spelsberg, T.C., Steggles, A.W. and O'Malley, B.W. 1971. Progesterone-binding components of chick oviduct. J. Biol. Chem., 246, 4188-4197.

Stumpf, C. 1965. Drug action on the electrical activity of the hippocampus. Int. Rev. Neurobiol., 8, 77-138.

Stumpf, W.E. 1968. Estradiol-concentrating neurons: Topography in the hypothalamus by dry mount autoradiography. Science, 162, 1001-1003.

Tindal, J.S., Knaggs, G.S. and Turvey, A. 1967. Central nervous control of prolactin secretion in the rabbit: Effect of local oestrogen implants in the amygdaloid complex. J. Endocrinol., 37, 279-287.

Toft, D. ad Gorski, J. 1966. A receptor molecule for estrogens: Isolation from the rat uterus and preliminary characterization. Proc. Nat. Acad. Sci., 55, 1574-1581.

Valenstein, E., Kakolewski, J.W. and Cox, V.C. 1967. Sex differences in task preference for glucose and sascharin solutions. Science, 156, 942-943.

Van Loon, G.R. and Ganong, W.F. 1971. Evidence for central adrenergic neural inhibition of ACTH secretion in the rat. Endocrinology, 89, 1464-1469.

Vardaris, R.M. and Schwartz, K.E. 1971. Retrograde amnesia for passive avoidance produced by stimulation of dorsal hippocampus. Physiol. Behav., 6, 131--135.

Wade, G. and Zucker, I. 1970. Modulation of food intake and loco-

motor activity in female rats by diencephalic hormone implants.
J. Comp. Physiol. Psychol., 72, 328-336.
Wagner, A.R. 1966. Frustration and punishment. In: Current
Research on Motivation, R.N. Haber, (Ed.). New York, Holt,
Rinehart, and Winston, pp. 229-239.
Weiss, J.M., McEwen, B.S., Silva, M.T. and Kalkut, M. 1970.
Pituitary-adrenal alterations and fear responding. Amer. J.
Physiol., 218, 864-868.
Zigmond, R.E. 1971. Chemical and anatomical specificity of
gonadal hormone retention in the rat brain. Ph.D. Thesis, The
Rockefeller University.
Zigmond, R.E. and McEwen, B.S. 1970. Selective retention of oes-
tradiol by cell nuclei in specific brain regions of the ovariec-
tomized rat. J. Neurochem., 17, 889-899.

SEXUAL MOTIVATION

Seymour Levine

Department of Psychiatry, Stanford University School of

Medicine, Stanford, California

There are certain obvious things which came out of the
sessions we heard both this morning and this afternoon and there
are also so many questions that it is impossible in the limited
amount of time to discuss them all.

But the one thing which is now certain is that the brain is
indeed a target organ for hormones. The brain is a receptor site
for hormones. It sees the hormones, integrates them and acts upon
whatever information they impart.

In the many years during which there has been a major thrust
in neuroendocrinology it has been essentially on how the central
nervous system controls hormones. What we are seeing now is the
other side of the coin--how do the hormones regulate the action of
the brain, how do they impart information to the brain upon which
the brain acts? But in order to have any kind of information system you need both your chemical and its receptor sites. How the
hormone acts upon these receptors is dependent on some state or
process of the brain itself, on some form of neural organization.
This has been very clearly demonstrated by Dr. Whalen in some of
his earlier work on sexual differentiation and some of our work
which shows that males and females respond very differently to
homotypical or heterotypical hormones. It is very easy to induce
a lordosis response with minimal amounts of estrogen and progesterone in the female but, although it is possible (difficult to
say the least) to produce a lordosis response in the male, and if
you produce it you do so using much larger amounts of a hormone
given over much longer periods of time, I have never yet seen data
which indicate a normal female lordosis-to-mount ratio no matter

what amount of estrogen you give the male. We can alter this
situation; if the testes are removed from the newborn male this
male becomes about as sensitive as the female and shows lordosis
behavior that is almost identical to that of the female. So
receptor sites in the male can be altered by depriving the male of
the normally occurring testosterone during critical periods of
development and leave the brain essentially in the receptor state
you would normally find the female. One can abolish this easily
by giving the male small amounts of testosterone at the time you
castrate so that you permit the appropriate neural organization to
occur to make the receptor sites now male.

But there is a problem which has always puzzled me and which
I think I'd like to address to Dr. Whalen, maybe to the whole
group. This is--we all like to talk in generalities and we seem
to neglect the fact that there are indeed marked individual differences. Taking the problem of castration and the continual
maintenance of sexual behavior, it has been reported by Knut Larsson
and others that there are marked individual differences. First,
there are marked individual differences in the sexual activity of
the normal male so that there is some percentage of males which
never show sexual behavior although, as far as we know, their
hormonal status is perfectly normal. There are some males we will
define as low copulators so that in a standard test situation
these males will tend to copulate with a very low degree of frequency. We have a range going up to very high copulators--males
that will have a large number of ejaculations in a fixed test
period. It has been demonstrated that if you castrate these males
and replace them with super threshold amounts of testosterone, all
you reinstate is the amount of behavior the male showed prior to
castration. If you look at the maintenance of the behavior you
will find that the high copulators tend to maintain their behavior
following castration much longer than the low copulators. In fact,
low copulators frequently show an almost immediate cessation of
sexual behavior following castration. Here again you have a neurally organized brain, a male organized brain, which still shows a
marked differential sensitivity in terms of the amount and quantity
of behavior elicitable under influence of the hormone. It isn't
that you can't change low copulators. Unfortunately, the only
conditions under which we have ever been able to change low copulators to high copulators are those which are non-hormonal. Thus,
while sleep deprivation (REM deprivation) will not alter a high
copulator--a high copulator continues to be a high copulator--it
will take a low copulator and make it into a high copulator. There
is a question I am going to ask Dr. Whalen which I would like him
to think about and perhaps respond to in regard to the problem of
anti-hormones. It was very clear in his work that the anti-estrogen had an effect on female sex behavior. What is puzzling is
why the anti-androgen does not seem to have an effect on adult male

sexual behavior. Although the anti-androgen appears to have a
very marked effect on gonadotropin secretion, as demonstrated by
Davidson and Bloch, for some reason it appears to have little
effect on male sex behavior; in fact there is some evidence that it
may enhance it.

I'd like now to move to Dr. Green whom, of all the people in
the world, I would like to believe. The only trouble is I come
away somewhat unconvinced and for one very good reason--the data
on the adrenogenital syndrome does not hold together. Here you
have high amounts of androgen secreted prenatally and yet, although
you have a masculinization as far as morphology is concerned, the
evidence on masculinization of these girls in adulthood is really
very minimal. This makes me terribly uncomfortable in terms of
making the kinds of generalities we would like to make from the rat
studies to the human. Clearly, this is the purest case of androgens
in high quantities being secreted in utero and yet the amount of
sexual change that occurs is really quite minimal.

I always find it difficult to respond to Bruce McEwen. His
data are so elegant that I sit in awe and wonder how I can possibly
discuss it, so what I'll do is, I won't discuss it but will move
within the vein of his work, point out something which I think
really needs to be pointed out. It is still pervasive, at least in
the literature and among the people who work in areas of the adrenocortical hormones, ACTH and behavior, that we are dealing essentially
with a stress-behavior notion and that these things happen under
conditions of stress. I think what Dr. McEwen may realize is that
in effect there is much stronger ground for his case. If you look
at the range of behaviors that have now been implicated with regard
to the adrenocortical or pituitary-adrenal system, it isn't only
conditioned avoidance response which seems to be influenced; it
isn't only those behaviors which are related to stress; in fact,
one can now cite a catalog of behaviors starting with simple processes of habituation and orientation which appear to be related to
ACTH and steroids. Endroczi has shown that EEG habituation to a
flashing light and to a sound is markedly retarded if ACTH or ACTH
fragments are given. In some of our own work with one of my ex-
graduate students, Joan Johnson, we showed that habituation to a
startle response (the response to a sudden onset of sound) is
markedly retarded in adrenalectomized animals. On the other hand,
if you take an implant of hydrocortisone and put it into the median
eminence of the hypothalamus, habituation is markedly accelerated.
We talked only about extinction, but there is now sufficient work,
again some done by Sheila Guth in our own laboratory, showing that
there are marked acquisition effects, and not acquisition effects
of typical aversive stimulation but acquisition effects of appetitive
stimuli, that are related to ACTH and adrenocortical hormones. The
extinction effects of ACTH are well known. They appear to be

related not only to those events that occur in terms of negative
or aversive stimuli but also with positive stimuli. The extinction of an appetitive response is also retarded to some extent by
giving ACTH.

Some recent work by Gary Coover and Larry Goldman in my
laboratory has shown that if you take an animal on an appetitive
schedule after it is well stabilized and put it on a non-reinforce-
ment, that is an extinction schedule, the steroids elevate markedly
following the extinction schedule. However, in our hands the
hippocampal animal does not show the reduction of the stress
response; he shows a perfectly adequate stress response in response
to ether or novelty. What the hippocampal animal does not do,
however, is show this elevation of steroids following extinction.
Accordingly, it also tends to show a much slower extinction rate.
The hippocampus, and all of the things the hippocampus seems to be
associated with, is one of the intriguing neural structures that
we are dealing with and although I still cannot find as adequate
a handle on it as I would like, the data that Dr. McEwen has
presented will be extremely valuable in our ways of looking at the
nature of hormonal and hippocampal or hormonal and brain function
relationships.

In general I'd like to thank all three of the speakers this
morning.

BRAIN CATECHOLAMINES, AFFECTIVE STATES AND MEMORY

Seymour Kety, M.D., Professor of Psychiatry

Harvard Medical School, Director of Psychiatric Research

Laboratories, Massachusetts General Hospital

Those who have been concerned with designs for the human brain have come to realize that it is not enough to incorporate mechanisms for the reception and processing of sensory information, its storage and retrieval, and the programming and regulation of motor output. Adaptive behavior, by its very nature, requires a system for the evaluation and sorting of environmental inputs, and the development of the most appropriate and effective responses in terms of the survival of the individual or the species.

Walter Cannon emphasized the crucial importance which the peripheral sympathetic nervous and endocrine systems played in that process, overlaying the more specific behavioral outputs and broadcasting more general information in a process which could anticipate, prepare for, and facilitate responses appropriate for survival-significant situations.

The central components of that system have been less well-defined, but in the past decade there has been remarkable progress made in delineating some of their anatomical pathways, describing the behavioral functions in which they appear to be involved, and suggesting some of the chemical substances which may serve as transmitters or modulators.

LOCALIZATION AND DISTRIBUTION

The existence of biogenic amines in the brain was demonstrated by Marthe Vogt (1954) and others in a previous decade and in the one just past, the localization of these in particular neurons and their axonal endings has been elegantly shown, and their pathways exhaus-

tively traced by a number of Swedish workers, notably Fuxe and
Dahlström (1965), who, with Hillarp (1966), adapted the histofluor-
escence technique of Eränkö (1956) to the central nervous system.
Fuxe (1965) has summarized the distribution of these monoamine
pathways. Those which contain serotonin have their cell bodies
concentrated in the midline raphe nuclei of the brainstem, the
axons of which pass rostrally to a considerable extent through the
medial forebrain bundle to innervate much of the limbic system,
cerebrum and cerebellum. The catecholamine-containing neurons are
of two types; those which produce dopamine are highly concentrated
in the substantia nigra and send their axons through the nigro-
striatal tract to end in the nucleus striatum; other dopamine containing pathways are beginning to be defined. Those which contain
norepinephrine in great abundance are laterally distributed in the
brainstem with a high concentration in the locus ceruleus. The
medial forebrain bundle also serves as a funnel through which a
dense segment of catecholamine containing fibres pass on their way
to the diencephalon and telencephalon. Fuxe and his coworkers (1968)
have more recently described dorsal and ventral bundles with more
specialized distributions and functions. Monoamine-containing axons
also pass from the brainstem down into the spinal cord.

By use of biochemical, physiological, pharmacological and
behavioral observations, a body of compelling evidence has been
acquired by many workers, which suggests a role for these pathways
and their amines in affective or emotional states, representing
adaptive responses to environmental inputs which have survival significance. It may be useful to summarize some of the evidence that
catecholamines act at synapses within the brain in the mediation
of such states, even though one recognizes that many other substances
are probably involved. Somewhat less extensive, but nonetheless
substantial evidence speaks for a role of serotonin in certain
behavioral states. It is very likely that by means of some inter-
play which is as yet only poorly defined, the catecholamines and
serotonin as well as other amines may function to determine a highly
specific biochemical milieu which affects widely distributed synaptic
functions in the mediation of particular affective or motivational
states.

AROUSAL

There is considerable evidence that norepinephrine is involved
in the mediation of this generalized state that is associated with,
and facilitates, recognition and exploration of novel and significant
stimuli. Much of the evidence is pharmacological, which has the
drawback that drugs have many effects in addition to the one for
which they have been used, so that one cannot be certain that the
most obvious or parsimonious interpretation of a drug effect is the

valid one.   Some unexpected and unknown action of the drug may
actually have produced the observed effects.   However, if a number
of different drugs with a common action are used and produce consistent effects, the likelihood of an alternative action common to
them all is considerably reduced.

Sedation has long been known to be associated with the administration of reserpine, an agent which causes depletion of numerous
biogenic amines from central synapses.   The observation by Carlsson
and coworkers (1957), however, that the sedative effects of reserpine
could be promptly reversed by DOPA, a precursor of catecholamines,
but not by a precursor of serotonin, suggested that it was the
depletion of catecholamines which was responsible for the sedation
of reserpine.

Amphetamine is well known to counteract drowsiness and increase
arousal in animals and man.   Among the actions of this drug in the
brain are several which would be expected to augment the concentration of catecholamines at central synapses, by favoring their release
from presynaptic endings, or inhibiting their synaptic inactivation
by blocking monoamine oxidase or by impairing their reuptake
(Glowinski and Axelrod, 1965).   The dextro-isomer has been shown
preferentially to block the reuptake of norepinephrine while both
the D and the L isomers have an equal inhibitory action on the
reuptake of dopamine.   The fact that the dextro-isomer of amphetamine has a comparably greater effect on arousal is compatible with
the hypothesis that norepinephrine may be especially involved in
that state (Taylor and Snyder, 1971).   A drug (cataprezan) which in
peripheral synapses behaves like a norepinephrine agonise also
increases alertness in animals.   On the other hand, sedation appears
to result from administration of α-methylparatyrosine which blocks
the synthesis of catecholamines.   Disulfiram, which, among other
actions, inhibits dopamine β-hydroxylase (the enzyme responsible
for the conversion of dopamine to norepinephrine), and depletes
norepinephrine, but not dopamine or serotonin in the brain, permits
locomotion, but inhibits arousal and exploratory behavior under
normal conditions.   It also prevents the increase in these behaviors
which is ordinarily induced by dextro-amphetamine.

In contra-distinction to these observations, a large number
of investigators who injected norepinephrine or epinephrine into
the ventricles reported sedation or somnolence rather than hyperarousal (Mandell and Spooner, 1968).   More careful regulation of
the dosage or concentration of infused norepinephrine has, however,
been shown independently by two groups (Segal and Mandell, 1970;
Herman, 1970) to result in increased locomotion and exploratory
behavior at low levels giving way to stupor and abolition of motor
activity when the concentration is increased.

STRESS

There is indirect evidence that the release of norepinephrine
at central synapses is associated with a variety of stresses,
including exposure to cold, severe exercise, fear or pain. A
significant reduction in norepinephrine levels in the brainstem and
other regions, compatible with a release of that amine, has been
reported in certain severely stressful states, and an increased
turnover of tritium-labelled norepinephrine has been reported even
where distress was not associated with a measurable decrease in
endogenous levels (Thierry et al., 1968). The brainstems of stressed
animals have been shown capable of an increased synthesis of norepinephrine in vitro.

Repeated electroconvulsive shock has been found to result in
elevated levels of norepinephrine and tyrosine-hydroxylase (the
limiting enzyme in catecholamine synthesis) as well as an increased
turnover of radioactive norepinephrine in brain. These changes,
compatible with an increased synthesis of norepinephrine persisted
for at least 24 hours after the last shock (Kety et al., 1967;
Musacchio, 1969).

RAGE

Sham rage induced by electrical stimulation of the amygdaloid
nucleus in cats has been found to result in a significant depletion
of norepinephrine from the telencephalon, and an increased turnover
of that amine. Drugs which potentiate such rage or fairly specifically block aggressive behavior usually have actions upon brain
catecholamines which would result in their increased or decreased
actions at central synapses, respectively (Reis & Gunne, 1965; Reis
& Fuxe, 1969).

SELF-STIMULATION

The "reward system" of Olds and Milner has been extensively
studied from the pharmacological and behavioral points of view, resulting in rather definitive evidence for the involvement of norepinephrine at crucial synapses within it (Stein and Wise, 1970). Mapping
of the areas within the brain which result in positive reinforcement
or self-stimulation by animals shows a high concentration of these
points within the ascending norepinephrine system of the brain. Self-
stimulation is also found with electrodes in the midline. One group
of investigators has reported evidence that the positive reinforcing points all are within the ventral norepinephrine system (Arbuthnott et al., 1971).

Drugs which interfere with the action of norepinephrine at
central synapses by preventing its synthesis (α-methylparatyrosine),

depleting it from nerve endings (reserpine), or blocking catecholamine receptors (chlorpromazine), all suppress self-stimulation.
On the other hand, drugs which enhance norepinephrine at central
synapses (amphetamine or a combination of a monoaminoxidase inhibitor and tetrabenezine) facilitate such behavior (Stein and Wise,
1970).

More recently, Stein and Wise have obtained evidence which has
quite specifically implicated norepinephrine in this form of appetitive behavior. Inhibitors of dopamine β-hydroxylase (disulfiram
or diethyl-dithiocarbamate) block the synthesis of norepinephrine
in the brain, but not of dopamine or serotonin. Such agents suppress self-stimulation and block the ability of amphetamine to
increase it. Normal self-stimulatory behavior can be restored in
such an animal by intraventricuoar injection of L-norepinephrine,
but not by dopamine or serotonin. After the infusion of norepinephrine has replenished the depleted stores of that amine, administration of amphetamine will once again facilitate self-stimulation.
Norepinephrine intraventricularly administered to a normal animal
will also potentiate self-stimulatory behavior (Stein and Wise,
1970).

Evidence has also been acquired which suggests the release of
norepinephrine at certain synapses in the course of self-stimulatng
behavior. Arbuthnott and coworkers (1971) have shown a more rapid
depletion of norepinephrine following α-methylparatyrosine in the
hypothalamus, preoptic areas and parts of the limbic system, but
not in the cerebral cortex. They found no change in dopamine levels
in the striate nucleus.

Using a push-pull cannula, Stein and Wise, having previously
labelled the endogenous stores of catecholamine in the brain with
tritiated norepinephrine, found a release of radioactive material
from the hypothalamus and the amygdala only in conjunction with
rewarding stimulation. Such releases were not found in the cortex
or thalamus. Parenteral administration of amphetamine which is
capable of potentiating self-stimulation also facilitated the release
of radioactivity from the amygdala. Although an increase in methyl-
ated metabolites of norepinephrine has been identified, the specific
release of norepinephrine rather than its metabolites from these
regions has not yet been conclusively shown, although there is little
doubt from the indirect evidence, that such a release occurs (Stein,
1972).

HUNGER AND THIRST

In addition to self-stimulation, other appetitive behaviors
have been examined with similarly compelling evidence for the specific involvement of norepinephrine in their mediation. Slangen and

Miller (1969) reported the induction of feeding in satiated rats
following micro-injection of L-norepinephrine into the perifornical
region of the anterior hypothalamus. Serotonin did not produce such
an effect and the effect from dopamine was weak and delayed, compatible with its conversion to norepinephrine. Further evidence was
adduced to specify the type of adrenergic receptor involved in this
type of feeding behavior. L-norepinephrine acts on both α and β
receptors induced eating, but a pure β-agonist (isoproterenol) had
no effect. An α-antagonist (phentolamine) blocked the effect of
norepinephrine while a β-antagonist (propanalanol) had no such effect. This series of experiments quite definitely implicates norepinephrine in this behavior and indicates that α receptors to norepinephrine are involved.

By similar types of experiments in the same region, Leibowitz
(1970) and others have shown that an action on β-adrenergic receptors
was involved in drinking behavior while activation of α-receptors,
which initiated eating, suppressed drinking, suggesting that these
behaviors represent the resultant actions of at least two reciprocal
systems.

CONDITIONED RESPONSES, MEMORY AND LEARNING

Most theories of memory have emphasized the requirement of
repetitive activation of a pathway in the progressive development of
a persistent change in the properties of its component synapses.
In 1963, Chamberlain, Halick and Gerard demonstrated that a postural
asymmetry induced by a unilateral cerebellar lesion induced, in
less than one hour, a persistent asymmetry in function at the level
of the lower motor neuron. In further experiments, Chamberlain,
Rothschild and Gerard adduced evidence that this process could be
facilitated by an agent which stimulated RNA synthesis and was
retarded by an agent which inhibited that process (1963). This
report lent strong experimental support to the concept that repeated
activation of a synapse induced a persistent facilitation in its
function and suggested that synthesis of RNA or protein was involved
in the process.

It is, however, possible that the adaptive efficacy of such a
mechanism is remarkably enhanced in more complex neuronal systems and
later evolutionary development by the accretion of special regulatory
mechanisms playing upon this primordial process in a selective manner.

It is not difficult to see how, as a result of selective pres-
sure, certain rudimentary adaptive responses to prevalent environmental stimuli could become genetically endowed: crude aversive
behavior to noxious or threatening stimuli and appetitive behavior
to stimuli associated with safety, relief or discomfort, food, water,
or mating. A species which did not possess such built-in-responses

would soon have become extinct.

The long time scale over which evolutionary processes occur, however, permits the development of genetically transmitted responses only to environmental threats or promises which have had survival significance over the long experience of the species and its fore-bears. A more powerful adaptation would reside in the ability of the individual to learn from its own experience what are threats, what environmental presentations have ultimate survival advantage, and what are the most effective patterns of behavior for defending against one or achieving the other. It is possible to suggest neuronal and neurochemical mechanisms which by constant reference to and interaction with a small number of inborn and primitive values, permits the nervous system to grow increasingly elaborate and effective in determining and preserving significant sensory patterns, appropriate evaluations, and effective behaviors all of which are based upon idiosyncratic experience. Only three such inborn states should be required: arousal, pain, and pleasure, including sensory patterns which regularly evoke them and appropriate neurogenic, humoral, visceral and motor responses to each. Arousal, although it would accompany both pain and pleasure, may also occur in the absence of either, i.e., in response to novel or unfamiliar stimuli and to inputs genetically at first and later, experientially, endowed with significance to the organism.

Thus, accompanying each of the relatively few environmental presentations to which were initially attached special significance for survival by the history of the species, there is an affective state consisting of a heightened level of arousal and the activation of particular circuits with a resultant drive to approach or to avoid the confronting stimulus. There are also induced a large number of appropriate peripheral autonomic discharges, all inborn and serving adaptive functions. If an additional mechanism existed whereby the emotional state accompanying the favorable or unfavor-able outcome of such behaviors could produce generalized intracere-bral release of a trophic neurogenic substance, an additional adaptation might be subserved, i.e., the transcription into more permanent form of the present and immediately preceding states of neuronal activation where the outcome had been significant to the organism.

Thus, the sounding of an unfamiliar tone produces a character-istic signal which, by virtue of its novelty, disinhibits and gains access to large numbers of neurons indiscriminately throughout the neocortex. After a very brief interval a painful shock is applied to the paw and the animal makes an aversive movement mediated by certain built-in reflexes. Another pathway, equally built-in, activates affect-mediating neurons which evoke the peripheral

autonomic responses associated with pain. Kety (1971) has suggested
that through other central pathways these same neurons produce a
generalized discharge about cortical synapses indiscriminately which
would include those still reverberating or otherwise sustaining a
residual response to the tone. If one effect of that discharge is
to convert such short lived activity to a more persistent increase
in synaptic conductivity, one can see how, by a process of algebraic
summation like that employed in the computers of average transients,
a synaptic association would, with repetition gradually develop,
between the shock and any consistently associated sensory input.
Eventually the tone itself, without the intervention of the shock,
would find facilitated pathways that would evoke the aversive
response and the affective state originally linked to the shock. A
conditioned avoidance and conditioned emotional response would have
been established.

It is possible to suggest certain anatomical pathways and neurochemical mechanisms which may satisfy some of the requirements of
this model or which seem to merit further investigation from that
point of view. Scheibel and Scheibel (1967) have described in
greater detail the architectonics of the "unspecific afferents" in
the cerebral cortex with evidence that these long climbing axons
weave about the apical dendrites of pyramidal cells with an extremely
loose axodendritic association in contrast to the vast number of
well defined synapses established by the terminals of the "specific
afferents." In 1968, Fuxe, Hamberger and Hökfelt described the
terminations of the norepinephrine-containing axons in the cortex,
pointing out the similarities in their distribution to that of the
unspecific afferents of Scheibel and Scheibel.

If some of the unspecific afferents are indeed "adrenergic" or
"aminergic" terminals invading the millions of sensory-sensory and
sensory-motor synapses of the cortex, they would provide a remarkably effective mechanism whereby amines released in emotional states
could affect a crucial population of synapses throughout the brain.
The hippocampal and cerebellar cortex are also characterized by
"climbing fibres" some of which are norepinephrine-containing (Anden
et al., 1967; Blakstad et al., 1967). Reivich and Glowinski (1967)
have mentioned the concentration of ^{3}H-labelled norepinephrine about
the apical dendrites of pyramidal cells in the hippocampal cortex.
One group has come close to extablishing a transmitter role for that
amine between the terminals of certain climbing fibers and Purkinje
cells (Siggins et al., 1969). There is even some evidence (Anden
et al., 1966) that axons from the same adrenergic neurons in the
brain stem may be distributed to cerebral, hippocampal and cerebellar
cortex as well as to hypothalamus and other areas.

Marr (1969) has proposed a novel theory of the cerebellar cortex
which implies a "learning" of patterns of motor activity by that

structure. It is possible that by similar and simultaneous processes in these three cortices an affective state may serve concurrently
to reinforce and consolidate new and significant sensory patterns in
the neocortex, their affective associations in the hippocampus and
the complex motor programs in the cerebellum, all of which were
temporally associated with a painful or pleasurable outcome. Fuxe
and Hanson (1967) have adduced evidence for an increased turnover
and presumably a release of norepinephrine throughout the brain
during the performance of a conditioned avoidance response. This
release was not attributable to the stress or the motor activity
involved.

Just as the peripheral expressions of arousal and affect are
mediated by both neurogenic and endocrine processes, it is possible
that the central components employ humoral as well as neural modes.
Some important adaptive function may be served by the tendency of
apical dendrites and their afferents, in a remarkably similar fashion
in the cortices of the cerebrum, hippocampus and cerebellum, to
seek the surface, and by the cortical convolutions which increase
that surface several-fold. Developmental history and the geometrical
requirements of the circuitry have been invoked to explain this
phenomenon, but it is also possible that the position of these
structures in close proximity to the cerebrospinal fluid permits an
exchange of important molecules between them. The constant flow
of this medium from the ventricles over the whole cortical surface
on its way to arachnoid villi offers a means of superfusing the cortex
with substances derived form the blood stream at the choroid plexus
or released by various stations along its path. It is noteworthy
that ^{3}H-norepinephrine injected into one ventrical or into the cis-
terna magna rapidly penetrates the superficial layers of the brain
(Reivich and Glowinski, 1967). Since the amines do not easily pass
the blood-brain barier and would not readily be removed by the
capillaries, such a mechanism would further assure the widespread
distribution any of which may be released from endings near the
cortical surface.

Secretions of the hypothalamus, trophic hormones of the pituitary and the steroid hormones of the adrenal cortex, some of which
are regularly secreted in states of arousal, anxiety and stress, may
thus have access to this rich population of synapses. Many of these
substances have, in one system or another, displayed a capacity to
affect the synthesis of RNA or of protein. Corticosterone has
recently been found to restore tryptophan hydroxylase activity in
the midbrain of the adrenalectomized rat (Azmitia and McEwen, 1969).
The steroid hormones and ACTH influence acquisition and extinction
in reciprocal ways (De Wied, 1969; Levine and Brush, 1967). It is
tempting to speculate upon the possibility that by their actions
on cortical synapses, consolidation may be delayed during the

anxiety of an unresolved situation or facilitated when a favorable
outcome is achieved and the blood corticosteroid level falls.

The suggestion that the release of catecholamines may favor
consolidation of learning by stimulating protein synthesis is made
more tenable by recently acquired information on the possible action
of cyclic-3´, 5´-AMP in the brain. This substance, present in
surprisingly high concentration in the central nervous system as is
adenyl cyclase, the enzyme which brings about its synthesis, is
crucially involved in enzyme induction and protein synthesis in a
wide variety of bacterial and mammalian cells and appears to increase
the activity of a protein kinase in brain (Miyamoto et al., 1969).
Evidence is accumulating that cyclic-3´, 5´-AMP may mediate the
effects of norepinephrine on central neurons (Siggins, Hoffer & Bloom,
1969) as it is believed to do for action of catecholamines and other
hormones on liver, muscle, and other peripheral tissues. It is interesting that the stimulation of protein kinase by cyclic AMP can be
markedly potentiated by magnesium or potassium ions and inhibited
by calcium which suggests means whereby an effect of adrenergic
stimulation could be differentially exerted on recently active or
inactive synapses.

In spite of the absence of ribosomes in presynaptic terminals
examined by electron microscopy, there is evidence for the occurrence
of protein synthesis at synapses and information which suggests the
possibility of catecholamine-stimulated synthesis. Droz and Barondes
(1969) have demonstrated the appearance of radioactive protein at
nerve endings in the mouse cerebrum within 15 minutes after the
intracerebral administration of tritiated amino acids and, more
recently, Barondes (personal communication) has found evidence for
a rapid incorporation of labelled glucosamine into synaptosomal
fractions of brain. It is possible that new synaptic membrane may
readily be formed from polypeptides and glucosamines present at nerve
endings. Synthesis of neuronal membrane may be stimulated in brain
tissue by norepinephrine. Hokin (1969) has recently reported that
this amine stimulates phospholipid production in brain slices, while
Snedden and Keen (1970) have found an increased incorporation of ^{32}P
into this major component of neuronal membrane in both synaptosomal
and mitochondrial brain fractions on exposure to norepinephrine at
relatively high concentrations.

This hypothesis would predict that drugs which release or
enhance norepinephrine in the brain or exert its neuronal effects
would favor consolidation and facilitate memory. Amphetamine and
caffeine appear capable of producing such effects (Bignami et al.,
1965; Oliverio, 1968) and recently a more specific ability of
amphetamine or footshock to counteract the suppression of consolidation brought about by cycloheximide has been reported (Barondes
and Cohen, 1968). Amphetamine is believed to increase the release

and persistence of norepinephrine at central synapses while
caffeine may potentiate the post-synaptic activity of that amine
by ehhancing cyclic AMP concentration through its effects on
phosphodiesterase. More recently, amphetamine has been found
markedly to accelerate consolidation of the postural asymmetry
induced in the spinal cord by cerebellar lesions (Palmer et al.,
1970).

Conversely, lesions or drugs which cause the depletion or
blockade of norepinephrine at synapses should retard consolidation.
There are many reports on the ability of such drugs as reserpine
or α-methyl tyrosine to block an established conditioned avoidance
response. Such evidence is more relevant to the expression of memory
than to its consolidation (Roberts et al., 1970).

On the other hand acquisition or consolidation of learning
appears to be influenced by agents which deplete the brain of amines,
or, more specifically, of the catecholamines or norepinephrine.
Reserpine injected immediately after training has been found to
impair acquisition (Dismukes, 1970). The same drug appears to
depress uptake of labelled uridine and to slow consolidation of
postural asymmetry in the spinal cord (Palmer et al., 1970). Essman
(personal communication) found a significant impairment in acquisition and retention following α-methyl tyrosine; these deficits were
greatest at the time when brian norepinephrine was lowest. More
recently, diethyldithiocarbamate, an inhibitor of dopamine-β-hydroxylase, which is capable of depleting the brain of norepinephrine
but not of serotonin or dopamine, has been found to produce a significant amnesia when administered 30 minutes before or immediately
after a training trial but not when the administration is delayed
for two hours (Randt et al., 1971).

Lesions of the medial forebrain bundle which would be expected
to deplete the telencephalon of norepinephrine, but also of serotonin
and dopamine, are reported in two studies to depress acquisition,
among other effects (McNew and Thompson, 1966; Sheard et al., 1967).
Conversely, stimulation of this region has been used as a positive
reinforcement in conditioning curarized animals (Trowill, 1967;
Miller and DiCara, 1967).

The experiments of Miller and DiCara (Miller and DiCara, 1967;
DiCara and Miller, 1968) illustrate a corollary of the proposed
model. Since the adrenergic axons to the telencephalon are funnel-
led through the medial forebrain bundle, stimulation there could be
positively reinforcing by causing the generalized release of nore-
penephrine at cortical synapses which is presumed to occur in affective states, and by facilitating acquisition through enhanced
consolidation. Furthermore, since this process should occur indis-

criminately throughout the brain to favor any contingent synaptic
activation, bizarre and nonadaptive responses such as vasodilation
of one ear and vasoconstriction of the other could with simple
repetitive reinforcement readily be established by external design,
by-passing the evolutionary and inbred systems for evaluating the
adaptiveness of a response.

Although the evidence which has been cited is compatible with
the hypothesis presented, crucial evidence in its support is not
available. Libet and Tosaka (1970) have reported evidence for a
heterosynaptic interaction in sympathetic ganglia in which brief
exposure to dopamine produces facilitation of muscarinic depolarization which persists for two hours or more and suggest "a long-
lasting metabolic and/or structural change in the postsynaptic
neuron induced by the initial action of the modulating transmitter,
dopamine." Such an effect of one of the catecholamines on central
nervous synapses, or evidence that some metabolic process, such as
protein synthesis, at cortical synapses was directly affected by
central adrenergic activity would more specifically support the
hypothesis.

MOOD IN MAN

Evidence for the operation of catecholamines in human mood is,
of course, more indirect and less conclusive than in the studies
of behavioral states in animals, but a surprising degree of paralel-
lism appears to exist. As in the case of self-stimulation in
animals, drugs which are capable of interfering with adrenergic
synapses (reserpine, α-methyltyrosine, chlorpromazine) tend to cause
depression of mood. In fact, reserpine in 10-15% of patients who
receive it in the treatment of hypertension causes a state which is
difficult to distinguish from endogenous depression. On the other
hand, anti-depressant or euphoriant drugs have in common the ability
to potentiate noradrenergic synapses (monoaminoxidase inhibitors,
imipramine, amphetamine, cocaine) (Schildkraut and Kety, 1967).
Although L-dopa, the precursor of dopamine and norepinephrine, is
hardly an effective agent in the treatment of depression generally,
it has been found regularly to induce hypomania in patients with
bipolar manic-depressive illness. Alpha-methylparatyrosine has
also been found to reduce manic states (Gershon et al., 1971). It
is especially interesting that the dextro-isomer of amphetamine
(which acts preferentially on norepinephrine synapses of the brain)
has many times the potency of l-amphetamine in producing euphoria.
All of these observations are compatible with an important role of
norepinephrine at central synapses in the regulation of mood.

Compatible with this hypothesis also is the ability of repeated
electroconvulsive shocks to elevate norepinephrine in the brain and
to facilitate its synthesis (Kety et al., 1967). In preliminary

findings, Schildkraut (personal communication) has observed a marked rise in the excretion of 3-methoxy-4-hydroxy-phenylglycol paralleling the clinical improvement in depression following the initiation of electroshock therapy. That metabolite of norepinephrine is produced in the brain and there is some evidence that its excretion in the urine is indicative of its central rather than peripheral metabolism.

Although evidence that norepinephrine or any other biogenic amine is a transmitter at central synapses is still incomplete and much remains to be done in defining its particular role at different synapses, the evidence that this biogenic amine, and others, play a crucial role in mediating such affective and adaptive states as arousal, mood, fear and rage, or such processes as conditioning, learning, motivation or appetitive behavior is remarkably compelling.

REFERENCES

Anden, N.E., Fuxe, F. and Larsson, K. 1966. Effect of large mesencephalic-diencephalic lesions on the noradrenaline, dopamine and 5-hydroxytryptamine neurons of the central nervous system. Experientia, 22, 842-847.

Anden, N.-E., Fuxe, F., and Ungerstedt, U. 1967. A quantititative study on the nigro-neostriatal dopamine neuron system in rat. Acta Physiol. Scand., 67, 306-312, 1967.

Arbuthnott, G., Fuxe, K., and Ungerstedt, U. 1971. Cerebral catecholamine turnover and self-stimulation behavior. Brain Res., 27, 406-413.

Azmitia, E.C., Jr. and McEwen, B.S. 1969. Corticosterone regulation of tryptophan hydroxylase in midbrain of the rat. Science, 166, 1274-1276.

Barondes, S. H. Personal communication.

Barondes, S.H. and Cohen, H.D. 1968. Arousal and the conversion of "short-term" to "long-term" memory. Proc. Nat. Acad. Sci., 61, 923-929.

Bignami, G., Robustelli, F., Janku, I. and Bovet, D. 1965. Psychopharmacologie, C.R. Acad. Sci. (Paris), 260, 4273-4278.

Blakstad, T., Fuxe, K. and Hökfelt, T. 1967. Noradrenaline nerve terminals in the hippocampal region of the rat and the guinea pig. Z. Zellforsch., 76, 463-473.

Carlsson, A., Lindqvist, M. and Magnusson, T. 1957. 3,4-dihydroxy-phenylalanine and 5-hydroxytryptophan as reserpine antagonists. Nature, 180, 1200.

Chamberlain, T.J., Rothschild, G.H. and Gerard, R.W. 1963. Drugs affecting RNA and learning. Proc. Nat. Acad. Sci. 49, 918-924.

Chamberlain, T.J., Halick, P. and Gerard, R.W. 1963. Fixation of experience in the rat spinal cord. J. Neurophysiology, 26, 662-673.

Dahlström, A. and Fuxe, K. 1965. Evidence for the existence of monoamine neurons in the central nervous system. II. Experimentally induced changes in the intraneuronal amine levels of bulbospinal neuron systems. Acta Physiol. Scand., 64 (suppl. 247), 5-36.

De Wied, D. 1969. Effects of peptide hormones on behavior. In: Frontiers in Neuroendocrinology, ed. by W.F. Ganong and L. Martini. Oxford University Press.

DiCara, L.V. and Miller, N.E. 1968. Instrument learning of vaso-motor responses by rats: Learning to respond differentially in the two ears. Science, 159, 1485-1486.

Dismukes, R.K. and Rake, A. Cited in Roberts et al. 1970.

Droz, B. and Barondes, S.H. 1969. Nerve endings: Rapid appearance of labelled protein shown by electron microscope radioautography. Science, 165, 1131-1133.

Eränkö, O. 1956. Histochemical demonstration of noradrenaline in the adrenal medulla of the hamster. J. Histochem. Cytochem., 4. 11-13.

Essman, W.B. Personal communication.

Fuxe, K. 1965. Evidence for the existence of monoamine neurons in the central nervous system. IV. The distribution of monoamine terminals in the central nervous system. Acta Physiol. Scand., 64 (Suppl. 247), 37-85.

Fuxe, K., Hamberger, B. and Hökfelt, T. 1968. Distribution of noradrenaline nerve terminals in cortical areas of the rat. Brain Res., 8, 125-131.

Fuxe, K. and Hanson, L.C.F. 1967. Central catecholamine neurons and conditioned avoidance behavior. Psychopharmacologia, 11, 439-447.

Gershon, E.S., Bunney, W.E., Jr., Goodwin, F.K., Murphy, D.L., Dunner, D.L. and Henry, G.M. 1971. Catecholamines and affective illness: Studies with L-DOPA and alpha-methyl-para-tyrosine. In: Ho, B.T. and McIsaac, W.M., Brain Chemistry and Mental Disease, Plenum Press, New York., pp. 135-162.

Glowinski, J. and Axelrod, J. 1965. The effect of drugs on the uptake, release and metabolism of H^3-norepinephrine in the rat brain. J. Pharmacol., 149, 43-49.

Herman, Z.S. 1970. The effects of noradrenaline on rat's behavior. Psychopharmacologia, 16, 369-374.

Hillarp, N.-A., Fuxe, K. and Dahlström, A. 1966. Demonstration and mapping of central neurons containing dopamine, noradrenaline, and 5-hydroxytryptamine and their reactions to psychopharmaca. Pharmacol. Rev., 18, 727-741.

Hokin, M.R. 1969. Effect of norepinephrine on ^{32}P incorporation into individual phosphatides in slices from different areas of the guinea pig brain. J. Neurochem., 16, 127-134.

Kety, S.S. 1971. The biogenic amines in the central nervous system: Their possible roles in arousal, emotion and learning. In: The Neurosciences: Second Study Program. Schmitt, F.O. (Ed.). The Rockefeller University Press, New York, pp. 324-336.

Kety, S.S., Javoy, F., Thierry, A.-M., Julou, L. and Glowinski, J. 1967. A sustained effect of electroconvulsive shock on the turn-

over of norepinephrine in the central nervous system of the rat.
Proc Nat Acad. Sci., 58, 1249–1254.

Leibowitz, S.F. 1970. Reciprocal hunger-related circuits involving
alpha- and beta-adrenergic receptors located, respectively, in
the ventromedial and lateral hypothalamus. Proc. Nat. Acad. Sci.,
67, 1063–1070.

Levine, S. and Brush, F.R. 1967. Adrenocortical activity and
avoidance learning as a function of time after avoidance training.
Physiol. Behav., 2, 385–388.

Libet, B. and Tosaka, T. 1970. Dopamine as a synaptic transmitter
and modulator in sympathetic ganglia: A different mode of
synaptic action. Proc. Nat. Acad. Sci., 67, 667–673.

Mandell, A.J. and Spooner, C.E. 1968. Psychochemical research
studies in man. Science, 162, 1442–1453.

Marr, D. 1969. A theory of cerebellar cortex. J. Physiol., 202,
437–470.

McNew, J.J. and Thompson, R. 1966. Role of the limbic system in
active and passive avoidance conditioning in the rat. J. Comp.
Physiol. Psychol., 61, 173–180.

Miller, N.E. and DiCara, L. 1967. Instrumental learning of heart
rate changes in curarized rats: shaping and specificity to dis-
criminative stimulus. J. Comp. Physiol. Psychol., 63, 12–19.

Miyamoto, E., Kuo, J.F. and Greengard, P. 1969. Cyclic nucleo-
tide-dependent protein kinase. III. Purification and properties
of adenosine 3′5′-monophosphate-dependent protein kinase from
bovine brain. J. Biol. Chem., 244, 6395–6408.

Musacchio, J.M., Julou, L., Kety, S.S. and Glowinski, J. 1969.
Increase in rat brain tyrosine hydroxylase activity produced by
electroconvulsive shock. Proc. Nat. Acad. Sci., 63, 1117–1119.

Oliverio, A. 1968. Neurohumoral systems and learning. In:
Psychopharmacology: A Review of Progress, 1957–1967, USPHS Pub.
No. 1836, U.S. Government Printing Office, pp. 867–878.

Palmer, G.C., Davenport, G.R. and Ward, J.W. 1970. The effect of
neurohumoral drugs on the fixation of spinal reflexes and the
incorporation of uridine into the spinal cord. Psychopharmacologia,
17, 59–69.

Randt, C.T., Quartermain, D., Goldstein, M. and Anagnoste, B. 1971.
Norepinephrine biosynthesis inhibition: Effects on memory in
mice. Science, 172, 498–499.

Reis, D.J. and Fuxe, K. 1969. Brain norepinephrine: Evidence that
neuronal release is essential for sham rage behavior following
brainstem transection in cat. Proc. Nat. Acad. Sci., 64, 108–112.

Reis, D.J. and Gunne, L. M. 1965. Brain catecholamines: Relation
to the defense reaction evoked by amygdaloid stimulation in cat.
Science, 149, 450–451.

Reivich, M. and Glowinski, J. 1967. An autoradiographic study of
the distribution of C^{14}-norepinephrine in the brain of the rat.
Brain, 90, 633–646.

Roberts, R.B., Flexner, J.B. and Flexner, L.B. 1970. Some evidence

for the involvement of adrenergic sites in the memory trace.
Proc. Nat. Acad. Sci., 66, 310-313.

Scheibel, M.E. and Scheibel, A.B. 1967. Structural organization
of nonspecific thalamic nuclei and their projection toward cortex.
Brain Res. 6, 60-94.

Schildkraut, J. Personal communication.

Schildkraut, J.J. and Kety, S.S. 1967. Biogenic amines and emotion.
Science, 156, 21-30.

Segal, D.S. and Mandell, A.J. 1970. Behavioral activation of rats
during intraventricular infusion of norepinephrine. Proc. Nat.
Acad. Sci., 66, 289-293.

Sheard, M.H., Appel, J.B. and Freedman, D.X. 1967. The effect of
central nervous system lesions on brain monoamines and behavior.
J. Psychiat. Res., 5, 237-242.

Siggins, G.R., Hoffer, B.J. and Bloom, F.E. 1969. Cyclic adenosine
monophosphate: Possible mediator for norepinephrine effects of
cerebellar Purkinje cells. Science, 165, 1018-1020.

Slangen, J.L. and Miller, N.E. 1969. Pharmacological tests for the
function of hypothalamic norepinephrine in eating behavior.
Physiol. Behav. 4, 543-552.

Snedden, J.M. and Keen, P. 1970. The effect of noradrenaline on
the incorporation of ^{32}P into brain phospholipids. Biochem.
Pharmacol. 19, 1297-1306, 1970.

Stein, L. 1972. Noradrenergic reward mechanisms, recovery of
function and Schizophrenia. In press, this volume.

Stein, L. and Wise, C.D. 1970. Mechanisms of the facilitating
effects of amphetamine on behavior. In D.H. Efron, (ed.),
Psychotomimetic Drugs, Raven Press, pp. 123-145.

Taylor, K.M. and Snyder, S.H. 1971. Differential effects of d-
and l-amphetamine on behavior and on catecholamine disposition in
dopamine and norepinephrine containing neurons of rat brain.
Brain Res., 28, 295-309.

Thierry, A.-M, Javoy, F., Glowinski, J. and Kety, S.S. 1968.
Effects of stress on the metabolism of norepinephrine, dopamine
and serotonin in the central nervous system of the rat. I. Modi-
fications of norepinephrine turnover. J. Pharmacol. Exp. Ther.,
163, 163-171.

Trowill, J.L. 1967. Instrumental conditioning of the heart rate
in the curarized rat. J. Comp. Physiol. Psychol., 63, 12-19.

Vogt, M. 1954. The concentration of sympathin in different parts
of the central nervous system under normal conditions and after
the administration of drugs. J. Physiol., 123, 451-481.

NORADRENERGIC REWARD MECHANISMS, RECOVERY OF FUNCTION, AND

SCHIZOPHRENIA

Larry Stein, C. David Wise and Barry D. Berger

Wyeth Laboratories, Philadelphia, Pennsylvania

". . . in its encounters with the environment the organism is guided by its responses of pleasure and pain. The former signals 'yes,' the latter 'no.' As we have seen, pleasure makes it move toward, and pain away from, the encountered stimulus. Pleasure then becomes the reward for successful performance, and the memory of pleasure incites repetition of the successful activity. Pain becomes the punishment for failure, and the memory of pain deters the organism from repeating the self-harming activity."

-- Sandor Rado, 1964

"The fact is decisive that the morbid anatomy [of dementia prae-cox] has disclosed not simple inadequacy of the nervous constitution but destructive morbid processes as the background of the clinical picture."

-- Emil Kraepelin, 1907

Over the past two decades, research findings in many different fields have resulted in the identification of central monoamine systems tentatively associated with reward and punishment. While the reward system has been characterized as mainly noradrenergic (Stein, 1967, 1968; Wise and Stein, 1969), the punishment system appears to be at least partially serotonergic (Wise, Berger and Stein, 1970, in press, a). Recent histochemical work suggests that nora-drenergic neurons may be organized into two ascending systems: a dorsal pathway originating in the locus coeruleus which mainly inner-vates the cerebral cortex and hippocampus, and a ventral pathway ori-ginating in the reticular formation of the lower brain stem which

mainly innervates the hypothalamus and ventral parts of the limbic
system (Fuxe, Hökfelt and Ungerstedt, 1970). Both pathways appear
to mediate rewarding effects (Ritter and Stein, 1972; Arbuthnott,
Fuxe and Ungerstedt, 1971), but their differential distribution suggests different functions: the ventral branch may mainly regulate
motivational activities, whereas the dorsal branch may mainly
regulate cognitive activities (Stein and Wise, 1971).

The cells of the serotonergic punishment system originate in
the raphe nuclei and distribute extensively in the central gray of
the midbrain; in addition, serotonin fibers ascend in the medial
forebrain bundle and distribute in the forebrain in roughly parallel fashion to those of the noradrenergic system. This innervation
may permit reciprocal control of behavior by norepinephrine and
serotonin neurons along lines generally suggested by Brodie and
Shore (1957), and illustrated diagrammatically in Figure 1.

Evidence in support of these conclusions has been reviewed
elsewhere (Olds, 1962; Stein, 1964a, 1968; Hoebel, 1971; Wise,
Berger and Stein, in press, a and b; Stein, Wise and Berger, in
press). Briefly, electrical stimulation of the medial forebrain
bundle serves as a powerful reward, and also elicits species-typical
consummatory responses, such as feeding and copulation, which produce pleasure and permit the satisfaction of basic needs. Similar
effects are produced pharmacologically by potentiation of the noradrenergic (or blockade of the serotonergic) activity of the medial
forebrain bundle. On the other hand, electrolytic lesions of the
medial forebrain bundle cause severe deficits in goal-directed
behavior and the loss of consummatory reactions; again, similar
effects are produced by pharmacological blockade of noradrenergic
function or potentiation of serotonergic function (Figures 2-5).
Finally, self-stimulation of the medial forebrain bundle increases
synthesis and turnover of norepinephrine (Wise and Stein, 1970;
Tables 1 and 2) and releases labeled norepinephrine into perfusates
of hypothalamus and amygdala (Stein and Wise, 1969). There is some
evidence that these findings in animals may be extrapolated to man
(Heath and Mickle, 1960; Sem-Jacobsen and Torkildsen, 1960).

How may these systems influence behavior, either explicit
(action) or implicit (thought). Briefly, our view is that they
provide feedback to ongoing behavior based on the consequences of
the behavior in the past (Mowrer, 1960; Stein, 1964b). The feedback
is positive in the case of reward or negative in the case of punishment. That is, one may think of the noradrenergic system in the
medial forebrain bundle as an action-facilitating (ventral pathway)
or thought-facilitating (dorsal pathway) mechanism. The mechanism
is activated or engaged by ongoing behavior if that behavior previously produced pleasure or avoided pain; such noradrenergic
activation initiates facilitatory feedback, which increases the

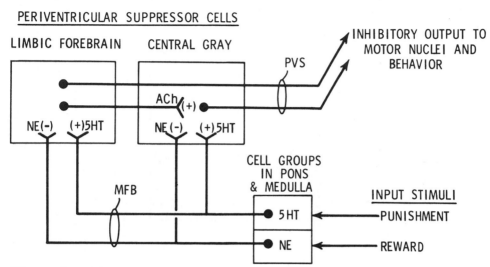

Figure 1.   Diagram representing hypothetical relationships between
reward and punishment mechanisms and behavior. Signals of positive reinforcement release behavior from periventricular system
(PVS) suppression by the following sequence of events: 1) Activation of norepinephrine-containing cells in lower brain stem by
stimuli previously associated with reward (or the avoidance of
punishment) causes release of norepinephrine (NE) into various
periventricular suppressor areas via the medial forebrain bundle
(MFB). 2) Inhibitory (-) action of NE decreases activity of PVS
suppressor cells in diencephalic and midbrain central gray either
directly, or by reducing their excitatory (+) input from cholinergic (ACh) suppressor cells in the limbic forebrain. 3) Reduction
of PVS inhibitory influence over motor nuclei of the brain stem
facilitates behavior. Signals of punishment or failure increase
behavioral suppression by activation of the dorsal raphe and other
serotonin (5HT) cell groups. Consequent release of 5HT either
excites PVS suppressor cells in the central gray directly, or indirectly via activation of cholinergic suppressor cells in the limbic
forebrain. See Dahlström and Fuxe (1965) for a detailed description
of the localization of norepinephrine and serotonin cell bodies in
the lower brain stem.

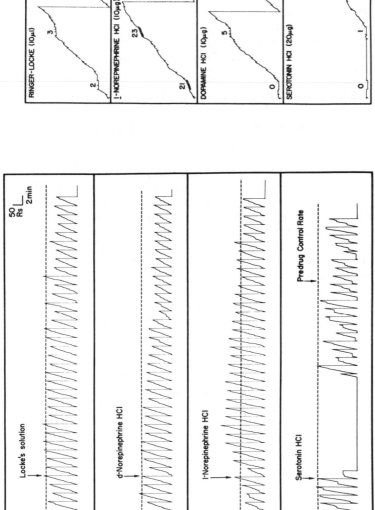

Figure 2. Facilitation of self-stimulation by norepinephrine and suppression by serotonin. The neurotransmitters (10 µg) were dissolved in 10 µl Ringer-Locke solution. Control injections of Ringer-Locke solution or d-norepinephrine (10 µg) had negligible effects. Pen cumulates self-stimulation responses and resets automatically every two minutes (see key). From Wise, Berger and Stein, in press (a).

Figure 3. Disinhibition of punished behavior by intraventricular administration of norepinephrine, and suppression of punished and unpunished behavior by serotonin (conflict test of Geller and Seifter, 1960). Each record represents a complete 72-minute session consisting of 4 three-minute punishment periods (punished lever-press responses are numbered) and 5 nonpunished periods of different durations, in which responses are reinforced at variable intervals (on the average, every two minutes). Pen resets automatically after about 550 responses. All drugs are dissolved in 10 µl of Ringer-Locke solution and injected in lateral ventricle via a permanently-indwelling cannula immediately before the start of the test. Disinhibitory effect of norepinephrine lasts for about 45 minutes, but suppressive effect of serotonin wears off after second punishment period.

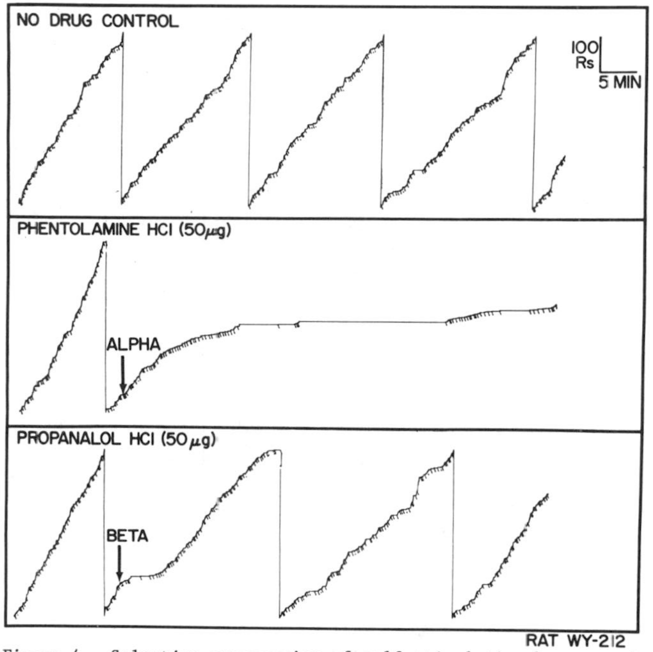

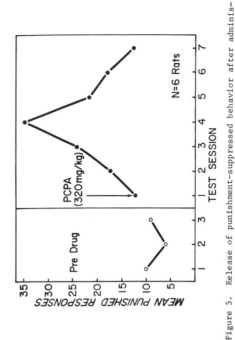

Figure 5. Release of punishment-suppressed behavior after administration of the serotonin synthesis inhibitor p-chlorophenylalanine (PCPA) (test of Geller and Seifter, 1960). PCPA was injected intraperitoneally on the day indicated, after completion of the behavioral test. The time courses of behavioral disinhibition and serotonin depletion after PCPA coincide closely. Administration of the serotonin precursor 5-hydroxytryptophan (5-HTP) reverses the disinhibitory effect of PCPA (not shown).

Figure 4. Selective suppression of self-stimulation by α-noradrenergic blockade. Phentolamine (α-antagonist) and propranolol (β-antagonist) were injected intraventricularly 15 min after the start of a 75-min test. Pen cumulates lever-press responses and resets automatically after 500 responses. Brain stimulus reinforcements were scheduled at variable intervals and, on the average, 15 seconds apart (indicated by pips on curves). From Wise, Berger and Stein, in press (a).

TABLE 1

Increased turnover of brain norepinephrine (NE) after 2 hours of
medial forebrain bundle self-stimulation. The rate of disappearance
of radioactivity from 6 brain regions was measured 18 hours after
an intraventricular injection of NE-C14 (7.6 µg, 3.2 x 10^6 dpm) in
6 self-stimulators and 6 implanted, but non-stimulated controls.
From Wise and Stein, 1970.

| Brain Region | Mean Radioactivity (dpm/mg) ± S.E.M. | | Mean % Difference |
	Control (N=6)	Self-Stimulation (N=6)	
Amygdala	11.7 ± 2.2	4.6 ± 1.1	−60.8*
Septum + Bed Nucleus of Stria terminalis	55.6 ± 7.9	25.3 ± 6.0	−54.6*
Hypothalamus	39.7 ± 4.4	20.6 ± 4.2	−48.2*
Hippocampus	18.8 ± 2.3	10.7 ± 1.4	−42.8*
Cortex	8.3 ± 1.4	5.0 ± 1.1	−40.2
Cerebellum	11.8 ± 1.3	8.3 ± 1.4	−29.2

*$P < 0.05$

TABLE 2

Increased turnover of brain norepinephrine (hypothalamus and limbic
system) after 3 hours of medial forebrain bundle self-stimulation.
Tyrosine-C14 (0.94 µg, 6.2 x 10^6 dpm) was injected intraventricularly
15 minutes before self-stimulation test.

| Treatment | No. of Rats | Specific Activity (dpm/µg) ± S.E.M. | |
		Norepinephrine	Tyrosine
Self-stimulation	4	1677 ± 249*	504 ± 87
Implanted Controls	4	2976 ± 398	732 ± 65

*Differs from control, $P < .05$.

probability that the behavior will run off to completion. On the
other hand, behavior associated with failure or punishment engages
the serotonergic punishment mechanism. This generates negative
feedback that tends to cut the unsuccessful behavior short. The
adaptive nature of these regulatory mechanisms is obvious. It is
equally obvious that malfunction of these mechanisms could have
serious consequences for normal behavior and logical thought. We
will argue that schizophrenia is one such extreme consequence,
and that it is caused by selective chemical damage to the noradrenergic reward mechanism.

ROLE OF HEREDITY IN SCHIZOPHRENIA

Genetic studies provide indirect support for the idea that an
impairment of noradrenergic function may be involved in schizophrenia. The early impression that schizophrenia is inherited has
been verified by systematic family studies (Kallman, 1938, 1946)
and studies of adopted children (Heston, 1970; Kety, Rosenthal,
Wender and Schulsinger, 1968), which establish "the importance of
genetic factors in the development of schizophrenia . . . beyond
reasonable dispute" (Heston, 1970). Several modes of inheritance
have been proposed, but some authorities currently favor the idea
that a main gene of large effect, modified either by a second gene
(Karlsson, 1966), or by multiple factors (Heston, 1970), is responsible for schizophrenia and borderline schizoid disorders. In
any case, the conclusion that schizophrenia is hereditary necessarily implies a biochemical aberration, since no other mechanism
is known for the expression of genetic traits.

A CHEMICAL THEORY OF SCHIZOPHRENIA

Recently, Stein and Wise (1971) proposed a novel physiological
and chemical etiology for schizophrenia. This work is based on
Thudichum's (1884) concept that "many forms of insanity" are
caused chemically by "poisons fermented within the body." The
essential properties of the offending chemical have been outlined
by Hollister (1968): "In short, what is required is an endogenous
toxin, highly active and highly specific in its action at minute
doses, continuously [or episodically] produced, for which tolerance
does not develop."

Current biochemical theories, which use mescaline or LSD as a
model, generally attribute hallucinogenic or psychotomimetic properties to the toxic metabolite (Friedhoff and van Winkle, 1962;
Hoffer and Osmund, 1959; Osmond and Smythies, 1952; Shulgin, Sargent
and Naranjo, 1969; Gottlieb, Frohman and Beckett, 1971). Such form-
ulations may be criticized on two grounds. First, in view of the
chronic and even life-long duration of schizophrenia, it seems
unlikely that a mescaline-like substance would be produced in

adequate quantities continuously over many decades without the
development of tolerance. Secondly, many authorities now question
the assumption that hallucinogenic drugs induce a "model psychosis."
These agents do not reproduce the fundamental symptoms of schizophrenia, and the differences between the drug states and schizophrenic reactions are easily distinguished (Hollister, 1968). The
wide variety of mental changes caused by the drugs, such as hallucinations and delusions, tend rather to resemble the accessory
symptoms of schizophrenia.

According to Bleuler (1950), "the fundamental symptoms consist of disturbances of association and affectivity." Specifically,
this author suggests that schizophrenic associations lose their
continuity because the thoughts

> . . . are not related and directed by any unifying
> concept of purpose or goal . . . . In analyzing
> the disturbances of association, we must realize the
> influences which actually guide our thinking. Asso-
> ciations formed in terms of habit, similarity, sub-
> ordination, causality, etc., of course will never
> generate truly fertile thoughts. Only the goal-directed
> concept can weld the links of the associative chain
> into logical thought . . . . The main objective
> [goal] will determine [the] associations . . . . The
> idea of water is quite different depending on whether
> it refers to chemistry, physiology, navigation, land-
> scape, inundation, or source of power . . . . No
> healthy person thinks of crystal water when his
> house is being swept away by a flood; nor will he
> think of water as a medium of transportion when he
> is thirsty.

Similarly, "schizophrenic behavior is marked by a lack of
interest, lack of initiative, lack of a definite goal, by inadequate
adaptation to the environment" (Bleuler, 1950); in short, schizophrenic behavior is inefficient (Wishner, 1965). At the same time,
emotional responsivity is diminished and eventually reduced to
indifference, so that "many schizophrenics in the later stages cease
to show any affect for years and even decades" (Bleuler, 1950). In
Rado's (1964) view, the disturbance of affect stems from the fact
that the "pleasure resources are inherently deficient," a view that
has been elaborated by Meehl (1962). Finally, Bleuler and others
emphasize that the course of the disease "is at times chronic, at
times marked by intermittent attacks, and which can stop or retrograde at any stage, but does not permit a full restitutio ad
integrum."

As noted previously, damage to noradrenergic reward mechanism

causes deficits in goal-directed behavior and rewarding consummatory
reactions in animals. By analogy, chemical damage of the reward
mechanism in the schizophrenic could produce deficits in goal-
directed thinking and behavior, and in the capacity to experience
pleasure. Furthermore, if the chemical toxin were continuously or
episodically produced, and if the damage to the reward mechanism
were at least partially irreversible, the disorder would be progressive and chronic.

POSSIBLE CHEMICAL CAUSE OF SCHIZOPHRENIA

Several lines of biochemical evidence suggested to us that
6-hydroxydopamine (2,4,5-trihydroxyphenethylamine) or a closely
related substance may be the aberrant metabolite that causes
schizophrenia (Stein and Wise, 1971). Like norepinephrine, 6-
hydroxydopamine is an oxidation product and metabolite of dopamine,
and "its formation can occur to a significant extent in the intact
animal" (Senoh et al., 1959). 6-Hydroxydopamine is difficult to
assay, since it is not only isomeric but also isographic with norepinephrine. Because of this similarity, 6-hydroxydopamine is
readily taken up into peripheral noradrenergic nerve terminals and
causes a marked and long-lasting depletion of norepinephrine
(Porter, Totaro and Stone, 1963). When injected intraventricularly
into the rat brain, 6-hydroxydopamine similarly causes a prolonged
or permanent depletion of brain catecholamines. Only catecholamine-
containing neurons are affected, and brain norepinephrine is more
severely depleted than dopamine (Uretsky and Iversen, 1969). Electron microscopic evidence reveals that norepinephrine nerve terminals in the brain degenerate and eventually disappear after repeated doses of 6-hydroxydopamine (Bloom et al., 1969).

Hypothalamic self-stimulation and other behaviors are markedly
suppressed for prolonged periods after single or repeated doses of
6-hydroxydopamine (Figure 6; Stein and Wise, 1971; Stein, 1971;
Breese, Howard and Leahy, 1971). Most significantly, the behavioral
deficits induced by 6-hydroxydopamine can be prevented by prior
administration of chlorpromazine, the drug of choice in the treatment of schizophrenia (Stein and Wise, 1971). At the same time,
chlorpromazine prevents the depletion of brain norepinephrine by
6-hydroxydopamine (Table 3). Finally, we and others have speculated that 6-hydroxydopamine may be a precursor both of a purported
schizophrenic sweat substance (Smith, Thompson and Koster, 1970)
and of possible hallucinogenic substances (Shulgin et al., 1969).

These considerations led us to propose the following etiology
of schizophrenia. As a result of a genetically determined enzymatic
deficit, 6-hydroxydopamine is episodically or continuously formed
in the schizophrenic brain and causes progressive damage to the
noradrenergic reward mechanism. In early stages, the deficits in

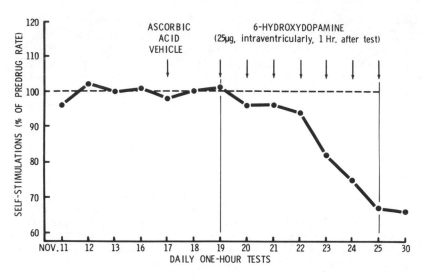

Figure 6.   Progressive suppression of self-stimulation by repeated doses of 6-hydroxydopamine (averaged data of 3 rats).   Lever-press responses were reinforced with electrical stimulation of the medial forebrain bundle at the level of the ventromedial hypothalamus. Note that the suppression of self-stimulation persisted for at least 5 days after dosing was discontinued.   From Stein and Wise, 1971.

reward function are largely reversible; hence, remissions are observed if the formation of 6-hydroxydopamine spontaneously ceases for an extended period, or if the chronic administration of chlorpromazine has blocked the uptake or production of the endogenous toxin.   In advanced stages, spontaneous recovery of normal behavior is quite rare and chlorpromazine is less efficacious, presumably because repeated exposure to 6-hydroxydopamine eventually causes irreversible damage to the reward mechanism.

SUPERSENSITIVITY TO NOREPINEPHRINE AND RECOVERY OF REWARD FUNCTION AFTER BRAIN DAMAGE

The model outlined above, which posits the deterioration of a specific neuronal system as the essential etiological factor in schizophrenia, is identical in its general concept to the early ideas of Kraepelin and Bleuler.   Later this concept became discredited, partly as a result of failures to find consistent patterns of morphological damage in the brains of schizophrenics (Dastur, 1959).   It is clear now, however, that specialized techniques, such as histochemical fluorescence and electronmicroscopy, would be necessary to detect the postulated damage to the noradrenergic system, which comprises only a small fraction of the neurons in the brain.

TABLE 3

Dose-related antagonism by chlorpromazine (CPZ) of the behavioral
and norepinephrine-depleting effects of 6-hydroxydopamine (6-HD).
Rats, injected daily with chlorpromazine HCl (3 or 10 mg/kg, i.p.)
or saline immediately after a 1-hr self-stimulation test, received
one hour later an intraventricular dose of 6-hydroxydopamine HCl
(25 μg) for 12 days. CPZ and saline injections were continued for
5 more weeks, during which time 3 larger doses of 6-HD (100, 100 and
200 μg) were administered at about weekly intervals. Norepinephrine
(NE) levels in limbic forebrain and hypothalamus were measured at
the end of the experiment.

Pre-treatment	No. of rats	Forebrain NE at end of experiment (% of untreated controls)	Mean self-stimulation rate (as % of control prior to 6-HD) during a series of 12 daily 25-μg dosings of 6-HD		
			Blocks of 4 Tests		
			1 - 4	5 - 8	9 - 12
Saline	4	3.8	65.4	55.3	53.4
CPZ (3mg/kg i.p.)	5	46.8*	91.7*	80.4*	80.0*
CPZ (10 mg/kg, i.p.)	4	75.1*	104.8*	100.9*	120.7*

* Differs from saline mean at P < .05.

    The frequent observation of remissions, especially in early
stages of schizophrenia, also seemed to some workers to be incom-
patible with the idea of permanent brain damage. This conclusion
is erroneous; indeed, the problem of recovery of function after
permanent brain damage is currently an active area of investigation.
For example, in animal studies, it is well established that electrolytic lesions of the lateral hypothalamus will abolish feeding
for several weeks. However, if starvation is prevented by forced
feedings, feeding gradually recovers even if most of the neurons
in the lateral hypothalamus are destroyed (Teitelbaum and Epstein,
1962).

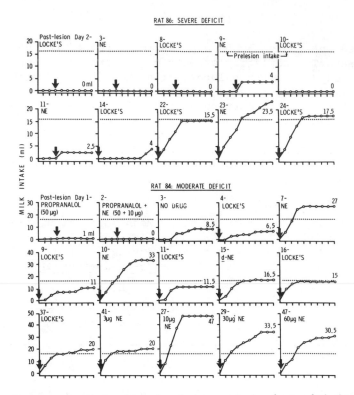

Figure 7. Recovery of feeding by intraventricular administration
of norepinephrine (NE) after lateral hypothalamic lesions. Curves
are cumulative plots of milk intake during 45-min tests in successive 5-min periods. Numbers over curves indicate 45-min intakes.
Arrows indicate time of intraventricular injections. All doses are
10 μg unless noted otherwise. (Rat 86) Reversal of aphagia by l-
norepinephrine (NE) in a rat with severe hypothalamic damage.
Postoperative feeding occurs for the first time on Day 9 as a
result of the norepinephrine injection, although a similar injection on Day 3 failed to induce feeding. (Rat 84) Reversal of
anorexia and extensive overeating induced by l-norepinephrine in a
case of moderate damage. On Day 15, d-norepinephrine had no effect.
The bottom row of plots indicates that norepinephrine is effective
over a wide range of doses, with optimal effect at 10 μg (From
Berger, Wise and Stein, 1971).

    In a recent study (Berger, Wise and Stein, 1971), we were able
to reverse the lateral hypothalamic anorexic syndrome by intraventricular administration of norepinephrine (Figure 7). This finding
suggested that the anorexia was due at least in part to a noradrenergic deficit; furthermore, since exogenous norepinephrine reversed
the syndrome, it seemed likely that noradrenergic receptors remained

functional and that the syndrome was due mainly to a deficiency of transmitter. We also observed that reversal of the lateral hypothalamic syndrome followed a definite time course. For the first several days after large bilateral lesions, norepinephrine failed to induce feeding. After a few weeks, however, total recovery of normal intake and even overeating could be induced by intraventricular norepinephrine. This time course is reminiscent of that observed in studies of disuse supersensitivity (Sharpless, 1969), and it suggests that such a mechanism (supersensitivity due to norepinephrine depletion) may explain the spontaneous recovery of feeding after hypothalamic damage. According to this idea, the gradual development of supersensitivity to norepinephrine after hypothalamic lesions compensates for the smaller amount of transmitter released from remaining undamaged neurons. Supersensitivity to norepinephrine may develop more readily than supersensitivity to acetylcholine in the lateral hypothalamic rat; if so, this could explain why feeding (a noradrenergically-mediated behavior) recovers rapidly after hypothalamic damage, whereas drinking (a cholinergically mediated behavior) recovers very slowly or not at all (Grossman, 1960, 1968; Teitelbaum and Epstein, 1962).

The recovery of self-stimulation and other behaviors after administration of 6-hydroxydopamine may depend on a similar process of denervation supersensitivity. Direct evidence of supersensitivity to norepinephrine after 6-hydroxydopamine in the self-stimulation test is shown in Figure 8: similar evidence has been reported by Mandell and coworkers (this volume) in tests of exploratory behavior. The increased responsivity to norepinephrine in our tests cannot be attributed to an artifact of the intraventricular injection procedure, since control rats injected with the ascorbate vehicle exhibited unchanged or diminished responses to repeated doses of norepinephrine. Neurohormonal specificity is suggested since self-stimulation was not facilitated by d-norepinephrine or dopamine. These findings leave little room for doubt that supersensitivity to norepinephrine is an important factor in the recovery of self-stimulation after 6-hydroxydopamine treatment.

Teitelbaum and coworkers (Teitelbaum, 1971; Teitelbaum, Cheng and Rozin, 1969) have demonstrated that the recovery of feeding after lateral hypothalamic damage parallels the development of feeding during maturation; all of the deficits in feeding and drinking observed during recovery from hypothalamic damage in adult rats may be observed in the developing patterns of feeding of infant rats. These workers speculated that the parallel depends on the fact that the same process of encephalization underlies both recovery and maturation. If so, and if the recovery of feeding after hypothalamic damage depends on the restoration of noradrenergic regulation, then maturation must depend on the development of noradrenergic regulation. We recently proposed a neurochemical scheme for

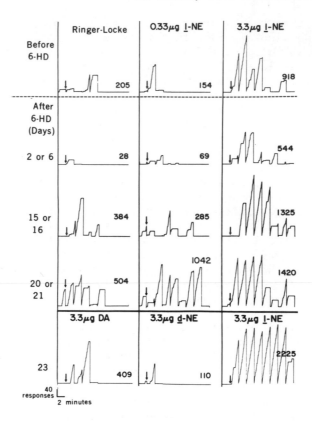

Figure 8. Development of supersensitivity to 1-norepinephrine
(1-NE) in a representative rat after 6-hydroxydopamine (6-HD)
treatment. A 100-μg dose of 6-hydroxydopamine was injected in the
lateral ventricle 2 hours after the self-stimulation test for two
successive days. Each record represents an 18-minute test and
shows the response output 2 minutes before and 16 minutes after
intraventricular injections of 1-NE, Ringer-Locke solution, dopamine
(DA), or d-norepinephrine (d-NE). Ringer-Locke solution and the
0.33-μg doses of 1-NE were always injected on the same day; the
3.3-μg doses of 1-NE were injected 1 or 4 days later. Self-stimulations are cumulated over time; the slope of the curve is proportion-
al to the response rate. Pen resets automatically every 2 minutes.
Numbers give total number of self-stimulations in the 16-minute
periods after injections (indicated by arrows). Note especially
dose-related facilitation of self-stimulation by 1-NE before 6-
hydroxydopamine, decreased responsivity to 1-NE 2 or 6 days after
6-hydroxydopamine, increased responsivity in the Ringer-Locke test
and supersensitivity to 1-NE 20 or 21 days after 6-hydroxydopamine,
and selective effect of 1-NE 23 days after 6-hydroxydopamine. Prior
to 6-hydroxydopamine treatment, the rate of self-stimulation was
stabilized at a low level by reduction of current intensity.

such regulation (Berger, Wise and Stein, 1971).

"Specifically, in the infant organism, the intake of
food is regulated by feeding centers in the lower
brainstem. During maturation, the feeding reflexes
come under the inhibitory control of suppressor cells
in the medial hypothalamus and the limbic forebrain.
The activity of these cells, in turn, is inhibited by
the noradrenergic food reward system in the lateral
hypothalamus. In the mature organism, appetizing foods
disinhibit the feeding reflexes from forebrain suppression by activating the noradrenergic food reward
system. Lateral hypothalamic damage stops feeding
because the forebrain suppressor cells no longer may
be inhibited in this way and, in fact, may be in re-
bound activation. Finally, during recovery from hypothalamic damage, the noradrenergic regulation of
suppressor cell activity is gradually and at least
partially restored."

By generalizing this concept, we can see that the developing
capacity for adaptive behavior (i.e., goal-directed action and
thought) may depend on the maturation of the noradrenergic reward
system. In other words, we assume with Teitelbaum (1971) that
behavioral regulatory mechanisms become increasingly encephalized
during maturation, and we further propose that such encephalization
is based in large measure on the gradually developing innervation
of higher regions of the brain by ascending noradrenergic reward
fibers in the medial borebrain bundle. It is easy to understand
from this hypothesis why damage to the medial forebrain bundle in
animals causes regression to primitive and less goal-directed modes
of behavioral regulation. It is our contention, furthermore, that
chemical damage to the noradrenergic reward system in man is responsible for the regressive patterns of behavior and thought in schizophrenia. Such a view is not inconsistent with the Freudian (1924)
concept that dementia praecox represents a regression to the earli-
est stage of psychosexual development.

Some support for these ideas is provided by developmental
studies of norepinephrine synthesis, which have been reviewed by
Kaufman and Friedman (1965):

"Karki et al. [1962] have shown that rats, 2 days
before birth, have essentially no norepinephrine in
their brain tissue. It did not reach the adult level
until six weeks after birth. Similar findings were
obtained with the rabbit. The guinea pig, however,
is born with high catecholamine levels. Since rats
and rabbits are essentially helpless at birth and

guinea pigs are almost self-sufficient, these findings
of biochemical immaturity agree with the behavioral
immaturity of these animals . . . .

These results imply that the enzymes involved in norepinephrine synthesis appear during development in
the same sequence as they operate in the norepinephrine
biosynthetic pathway in adult tissue. They also suggest
that behavioral maturity may be related to the development of pathways mediated by catecholamines."

Also interesting in this regard are observations that the deposition of neuromelanin (a pigment in catecholamine cells in the
brain) begins only after birth during the first five years of life,
and increases steadily throughout childhood and adolescence
(Fenichel and Bazelon, 1968). Interest in neuromelanin has been
stimulated by the suggestion that the pigment may be a marker of
catecholamine activity (Bazelon, Fenichel and Randall, 1967;
Cotzias et al., 1964).

Finally, it may be noted that the development of supersensitivity to norepinephrine in schizophrenia may provide an explanation
for certain symptoms, e.g., euphoria, manic and agitated states,
delusions of grandeur, etc., that are frequently observed in early
stages of the disease. In the context of our neurochemical model
these symptoms might seem to be paradoxical, since they would appear
to derive from an excess, rather than from a deficiency, of reward
activity. However, if moderate damage to noradrenergic terminals
in the early stages of schizophrenia induced supersensitivity to
norepinephrine, the powerful enhancement of noradrenergic function
that would result could cause exaggerated reward reactions during
later periods in which the formation of 6-hydroxydopamine had
ceased.

POSSIBLE ENZYMATIC FORMATION OF 6-HYDROXYDOPAMINE

Stein and Wise (1971) left open the question of the precise
mechanism of formation of 6-hydroxydopamine in the schizophrenic
brain. It was generally proposed that an error in the enzymatic
regulation of norepinephrine metabolism caused accumulation of
excess dopamine in the noradrenergic vesicle and the consequent
formation of 6-hydroxydopamine by autoxidation or an enzymatic
reaction. In addition, it was speculated that the enzymatic error
might involve dopamine-β-hydroxylase, the enzyme responsible for
the conversion of dopamine to norepinephrine.

Adams (in press) extended these ideas by proposing a "concrete
pathway for formation of endogenous 6-hydroxydopamine based on an
established enzymatic reaction." In this scheme, dopamine is

enzymatically converted by dopamine-β-hydroxylase to dopamine o-
quinone (DOQ), which in turn readily forms 6-hydroxydopamine if
intra-cyclization is blocked. Adams has kindly given us permission
to quote in detail from his forthcoming publication:

> "Thus, aberrant production of 6-OHDA [6-hydroxydopamine]
> can be explained if one has a rational, biochemical
> pathway leading to formation of small amounts of
> dopamine o-quinone (DOQ). To be relevant to the Stein
> and Wise postulate, this DOQ formation should at least
> satisfy the following four criteria: 1) be localized
> near NE nerve terminals where 6-OHDA formation has
> significance; 2) be small in magnitude since excessive
> formation of 6-OHDA would lead to severe destruction
> of NE nerve terminals; 3) be capable of sustaining
> itself over lengthy periods, yet probably more active
> under highly stressful conditions, consistent with the
> general clinical time pattern of schizophrenic illness;
> 4) be intimately associated with dysfunction of the
> dopamine-β-hydroxylase step envisaged by Stein and
> Wise, thus maintaining the concept of genetic predis-
> position to the biochemical aberration.
>
> Such a mechanism for DOQ production is indeed already
> well established in the literature. Kaufman and co-
> workers showed that dopamine-β-hydroxylase operates
> as a typical mixed-function oxidase and the normal
> hydroxylation of dopamine to NE is coupled to a
> stoichiometric oxidation of ascorbic acid as in the
> equation:

$$\text{Dopamine + Ascorbic Acid + O}_2 \xrightarrow{\text{enzymatic}} \text{NE + dehydroascorbic acid + H}_2\text{O}$$

> Levin and Kaufman (1961) showed that in the absence
> of ascorbate as electron donor, a small but significant
> conversion of dopamine to NE still occurred. They
> demonstrated clearly that in this case an "extra"
> molecule of dopamine served as reducing agent in
> place of ascorbate according to the reactions:

$$2 \text{ Dopamine + O}_2 \xrightarrow{\text{enzymatic}} \text{NE + DOQ + H}_2\text{O}$$

> The DOQ formed is re-reduced by a non-enzymatic reaction
> with DNPH:

$$\text{DOQ + DPNH + H}^+ \xrightarrow{\hspace{2cm}} \text{Dopamine + DPN}^+$$

> giving the net reaction:

$$\text{Dopamine} + O_2 + \text{DPNH} + H^+ \longrightarrow NE + DPN^+ + H_2O$$

Actual spectral evidence consistent with the enzyme-
catalysed dopamine o-quinone formation in the absence
of ascorbate and DPNH was obtained by Levin and Kaufman
(1961). Kaufman and Friedman (1965) reviewed the
details of these reactions.

There is little question but that DOQ will be formed
in the dopamine to norepinephrine conversion if a)
ascorbate is absent or very low in concentration,
or b) there is a dysfunction in the enzyme dopamine-
β-hydroxylase which causes it to utilize improperly
the available ascorbate."

The metabolic error in schizophrenia, according to Adams' proposal,
thus involves either excessive production of DOQ and/or a deficiency
in the capacity to recycle DOQ back to dopamine or otherwise dis-
pose of it.

We find much merit in this proposal. First, and most important, it provides a known enzymatic pathway for the formation of 6-
hydroxydopamine. Second, because dopamine-β-hydroxylase is required
for the formation of 6-hydroxydopamine, it plausibly explains why
the damage in schizophrenia might be restricted exclusively to
noradrenergic neurons. Dopaminergic neurons, which do not contain
dopamine-β-hydroxylase, would be spared. Such an account is required because schizophrenic patients ordinarily do not exhibit
Parkinson-like symptoms. Third, this proposal provides a detailed
basis for our suggestion that chlorpromazine may in part exert its
antipsychotic effect by blocking the entry of dopamine into the noradrenergic vesicle and thus reducing the formation of 6-hydroxydopa-
mine (Stein and Wise, 1972). Since dopamine-β-hydroxylase is
contained within the vesicle, it is evident that dopamine would
have to be taken up into the vesicle prior to its enzymatic conversion to 6-hydroxydopamine. Finally, because Adam's proposal pin-
points a specific dysfunction of dopamine-β-hydroxylase (that is,
utilization of dopamine as a reducing agent in place of ascorbate),
it increases the testability of our model.

POSSIBLE MODE OF DESTRUCTIVE ACTION OF 6-HYDROXYDOPAMINE
AND DOPAMINE o-QUINONE

Saner and Thoenen (1971) have proposed a plausible mechanism
to explain the destructive action of 6-hydroxydopamine. According
to this proposal, oxidation products of 6-hydroxydopamine (particularly, its p-quinone) form covalent bonds with brain protein and
thus render them nonfunctional. The reaction as such is nonspecific; selective destruction of noradrenergic neurons by exogenous

6-hydroxydopamine depends on the efficient uptake of the toxic substance into noradrenergic terminals. Similarly, we assume that the selectivity of the destructive effect of 6-hydroxydopamine in schizophrenia depends on the localized formation of the toxin in noradrenergic neurons.

Saner and Thoenen (1971) also indicate that "the o-quinone formed from DA [dopamine] is at least as reactive as the p-quinone derivative of 6-OH-DA [6-hydroxydopamine]." If so, then dopamine o-quinone (DOQ) could itself be the important toxic agent in schizophrenia. If DOQ also leads to the formation of 6-hydroxydopamine in the schizophrenic brain, then either DOQ or 6-hydroxydopamine, or both, could be causative agents in schizophrenia.

As indicated earlier, the pigmentation of the locus coeruleus and other catecholamine cell groups in man increases with advancing age. This brain pigment, neuromelanin, can be distinguished histochemically from melanin in the skin (Lillie, 1965). Since the skin pigment is formed from DOPA o-quinone, and since a dark pigment is formed enzymatically in an extract of brain from dopamine (but not from DOPA or epinephrine, and only to a small extent from norepinephrine; Wende and Spoerlein, 1963), it is reasonable to speculate that the brain pigment is formed from dopamine o-quinone. If so, it is not inconceivable that the conversion of DOQ to neuromelanin in the normal brain may be an important protective pathway, which may serve to inactivate DOQ and to prevent the formation of 6-hydroxydopamine. A defect in this pathway, or its overloading by excessive formation of DOQ, might be a factor in the etiology of schizophrenia. In this regard, it is of interest that chlorpromazine and other phenothiazine derivatives increase melanogenesis in peripheral tissue (Cotzias et al., 1964; Greiner and Nicolson, 1965). Such action could contribute to the antischizophrenic activity of the phenothiazines, if these drugs increased the deposition of neuromelanin in the locus coeruleus and other norepinephrine cell groups in the brain.

ACKNOWLEDGEMENTS

We thank Dr. Ralph N. Adams for stimulating discussions and for permission to quote from his forthcoming paper (Adams, in press) on the possible enzymatic formation of 6-hydroxydopamine.

Alfred T. Shropshire, Nicholas S. Buonato, William J. Carmint, Helen C. Goldman, John Monahan, Herman Morris, Janet D. Noblitt and Lois E. Wehren provided expert technical assistance.

L. STEIN, C.D. WISE AND B.D. BERGER

REFERENCES

Adams, R. N. 1972. Stein and Wise theory of schizophrenia: A
  possible mechanism for 6-hydroxydopamine formation in vivo.
  Behav. Biol., in press.
Arbuthnott, G., Fuxe, K. and Ungerstedt, U. 1971. Central cate-
  cholamine turnover and self-stimulation behavior. Brain Res.,
  27, 406-413.
Bazelon, M., Fenichel, G. M. and Randall, J. 1967. Studies on
  neuromelanin I. A melanin system in the human adult brainstem.
  Neurology, 17, 512-519.
Berger, B. D., Wise, C. D. and Stein, L. 1971. Norepinephrine:
  Reversal of anorexia in rats with lateral hypothalamic damage.
  Science, 172, 281-284.
Bleuler, E. 1950. Dementia Praecox, or the Group of Schizophre-
  nias, International Universities Press, New York.
Bloom, F. E., Algeri, S., Groppetti, A., Revuelta, A. and Costa,
  E. 1969. Lesions of central norepinephrine terminals with 6-
  OH-dopamine: Biochemistry and fine structure. Science, 166,
  1284-1286.
Breese, G. R., Howard, J. L. and Leahy, J. P. 1971. Effect of
  6-hydroxydopamine on electrical self-stimulation of the brain.
  Brit. J. Pharmacol., 43, 252-257.
Brodie, B. B., and Shore, P. A. 1957. A concept for a role of
  serotonin and norepinephrine as chemical mediators in the brain.
  Ann. N. Y. Acad. Sci., 66, 631-642.
Cotzias, G. C., Papavasiliou, P. S., Van Woert, M. H. and Sakamoto,
  A. 1964. Melanogenesis and extrapyramidal diseases. Fed. Proc.,
  23, 713-718.
Dahlström, A. and Fuxe, K. 1965. Evidence for the existence of
  monoamine-containing neurons in the central nervous system. I.
  Demonstration of monoamines in the cell bodies of brain stem
  neurons. Acta Physiol. Scand. 62, Supp. 232, 1-55.
Dastur, D. K. 1959. The pathology of schizophrenia. A.M.A. Arch.
  Neurol. Psychiat., 81, 83-96.
Fenichel, G. M. and Bazelon, M. 1968. Studies on neuromelanin
  II. Melanin in the brainstems of infants and children. Neurology,
  18, 817-820.
Freud, S. 1924. A General Introduction to Psychoanalysis (Trans.
  by J. Riviere) Boni & Liveright, London.
Friedhoff, A. J. and van Winkle, E. 1962. Isolation and charac-
  terization of a compound from the urine of schizophrenics. Nature,
  194, 897-899.
Fuxe, K., Hökfelt, T., and Ungerstedt, U. 1970. Morphological and
  functional aspects of central monoamine neurons. Int. Rev.
  Neurobiol., 13, 93-126.
Geller, I. and Seifter, J. 1960. The effects of meprobamate,
  barbiturates, d-amphetamine and promazine on experimentally induced conflict in the rat. Psychopharmacologia (Berlin) 1, 482-492.

Gottlieb, J. S., Frohman, C. E. and Beckett, P. G. S. 1971. The current status of the α-2-globulin in schizophrenia. In Biochemistry, Schizophrenias and Affective Illness, Himwich, H.E. (Ed.), Williams and Wilkins, Baltimore, pp. 153-170.

Greiner, A. C. and Nicolson, G. A. 1965. Schizophrenia-Melanosis. Lancet, 2, 1165-1167.

Grossman, S. P. 1960. Eating or drinking elicited by direct adrenergic or cholinergic stimulation of hypothalamus. Science, 132, 301-302.

Grossman, S.P. 1968. Hypothalamic and limbic influences on food intake. Fed. Proc., 27, 1349-1360.

Heath, R. G. and Mickle, W.A. 1960. Evaluation of seven years' experience with depth electrode studies in human patients. In Electrical Studies on the Unanesthetized Brain, Ramey, E.R. and O'Doherty, D.S. (Eds.), p. 214, Hoeber, New York.

Heston, L.L. 1970. The genetics of schizophrenic and schizoid disease. Science, 167, 249-256.

Hoebel, B. G. 1971. Feeding: Neural control of intake. Ann. Rev. Physiol., 33, 533-568.

Hoffer, A. and Osmund, H. 1959. The adrenochrome model and schizophrenia. J. Nerv. Ment. Dis., 128, 18-35.

Hollister, L.E. 1968. Chemical Psychoses. Thomas, Springfield, Ill.

Kallman, F. J. 1938. The Genetics of Schizophrenia. Augustin, New York.

Kallman, F. J. 1946. The genetic theory of schizophrenia: an analysis of 691 schizophrenic twin index families. Am. J. Psychiat., 103, 309-322.

Karki, N., Kuntzman, R., and Brodie, B. B. 1962. Storage, synthesis and metabolism of monoamines in the developing brain. J. Neurochem., 9, 53-58.

Karlsson, J.L. 1966. The Biological Basis of Schizophrenia. Thomas, Springfield, Ill.

Kaufman, S. and Friedman, S. 1965. Dopamine-β-hydroxylase. Pharmacol. Rev., 17, 71-100.

Kety, S.S., Rosenthal, D., Wender, P.H. and Schulsinger, F. 1968. The types and prevalence of mental illness in the biological and adoptive families of adopted schizophrenics. In The Transmission of Schizophrenia, Rosenthal, D. and Kety, S.S. (Eds.), pp. 345-362, Pergamon Press, New York.

Kraepelin, E. 1907. Introduction á la psychiatrie clinique, translated from the second German edition by A. Devaux and P. Merklen. Vigot freves, Paris.

Levin, E. Y. and Kaufman, S. 1961. Studies on the enzyme catalyzing the conversion of 3,4-dihydroxyphenethylamine to norepinephrine. J. Biol. Chem., 236, 2043-2049.

Lillie, R.D. 1965. Histopathologic technic and practical histochemistry. McGraw-Hill, New York.

Mandell, A.J., Segal, D.S., Kuczenski, R.T. and Knapp, S. 1972.

Some macromolecular mechanisms in CNS neurotransmitter pharmacology and their psychobiological organization. This volume.

Meehl, P. E. 1962. Schizotaxia, schizotypy, schizophrenia. Am.
Psychol., 17, 827-838.

Mowrer, O.H. 1960. Learning Theory and Behavior. Wiley, New York.

Olds, J. 1962. Hypothalamic substrates of reward. Physiol. Revs.,
42, 554-604.

Osmond, H. and Smythies, J.R. 1952. Schizophrenia: A new
approach. J. Ment. Sci., 98, 309-315.

Porter, C.C., Totaro, J.A., and Stone, C.A. 1963. Effect of 6-
hydroxydopamine and some other compounds on the concentration of
norepinephrine in the hearts of mice. J. Pharmacol. Exptl.
Therap., 140, 308-316.

Rado, S. 1964. Hedonic self-regulation of the organism. In The
Role of Pleasure in Behavior, Heath, R.D. (Ed.), p. 257, Hoeber,
New York.

Ritter, S. and Stein, L. 1972. Self-stimulation of the locus
coeruleus. Fed. Proc., 31, 820.

Saner, A. and Thoenen, H. 1971. Contributions to the molecular
mechanism of action of 6-hydroxydopamine. In 6-Hydroxydopamine
and Catecholamine Neurons, Malmfors, T. and Thoenen, H. (Eds.),
pp. 265-275, North-Holland Publishing Co., Amsterdam.

Sem-Jacobsen, C.W. and Torkildsen, A. 1960. Depth recording and
electrical stimulation in the human brain. In Electrical
Studies on the Unanesthetized Brain, Ramey, E.R. and O'Doherty,
D.S. (Eds.), p. 275, Hoeber, New York.

Senoh, S., Creveling, C.R., Udenfriend, S. and Witkop, B. 1959.
Chemical enzymatic and metabolic studies on the mechanism of
oxidation of dopamine. J. Am. Chem. Soc., 81, 6236-6240.

Sharpless, S.K. 1969. Isolated and deafferented neurons: Disuse
supersensitivity. In Basic Mechanisms of the Epilepsies.
Jasper, H.H., Ward, Jr., A.A. and Pope, A. (Eds.). pp. 329-348,
Little, Brown, Boston.

Shulgin, S.T., Sargent, T. and Naranjo, C. 1969. Structure-
activity relationships of one-ring psychotomimetics. Nature,
221, 537-541.

Smith, K., Thompson, G.F., and Koster, H.D. 1969. Sweat in
schizophrenic patients: Identification of the odorous substance.
Science, 166, 398-400.

Stein, L. 1964a. Self-stimulation of the brain and the central
stimulation action of amphetamine. Fed. Proc., 23, 836-850.

Stein, L. 1964b. Reciprocal action of reward and punishment
mechanisms. In The Role of Pleasure in Behavior, Heath, R.G.
(Ed.), pp. 113-139, Hoeber, New York.

Stein, L. 1967. Psychopharmacological substrates of mental depression. In Antidepressant Drugs, Garattini, S. and Dukes, N.M.G.
(Eds.), p. 130-140. Excerpta Medica Foundation, Amsterdam.

Stein, L. 1968. Chemistry of reward and punishment. In Psycho-

pharmacology: A Review of Progress, 1957-1967, Efron, D. H.
(Ed.), pp. 105-123, U.S. Government Printing Office, Washington.
Stein, L. 1971. Neurochemistry of reward and punishment: some
implications for the etiology of schizophrenia. J. Psychiat.
Res., 8, 345-361.
Stein, L. and Wise, C.D. 1969. Release of norepinephrine from
hypothalamus and amygdala by rewarding medial forebrain bundle
stimulation and amphetamine. J. Comp. Physiol. Psychol., 67,
189-198.
Stein, L. and Wise, C. D. 1971. Possible etiology of schizophrenia: Progressive damage of the noradrenergic reward mechanism
by 6-hydroxydopamine. Science, 171, 1032-1036.
Stein, L. and Wise, C. D. 1972. 6-Hydroxydopamine, noradrenergic
reward, and schizophrenia. Science, 175, 922-923.
Stein, L., Wise, C.D. and Berger, B. D. Antianxiety action of
benzodiazepines: Decrease in activity of serotonin neurons in
the punishment system. In Benzodiazepines. Garattini, S. (Ed.),
Raven Press, New York, In press.
Teitelbaum, P. 1971. The encephalization of hunger. In Progress
in Physiological Psychology, Stellar, E. and Sprague, J.M. (Eds.),
Academic Press, New York, pp. 319-350.
Teitelbaum, P., Cheng, M.F. and Rozin, P. 1969. Development of
feeding parallels its recovery after hypothalamic damage. J.
Comp. Physiol. Psychol., 67, 430-441.
Teitelbaum, P. and Epstein, A.N. 1962. The lateral hypothalamic
syndrome: Recovery of feeding and drinking after lateral hypo-
thalamic lesions. Psychol. Rev., 69, 74-90.
Thudichum, J.W.L. 1884. A Treatise on the Chemical Constitution
of the Brain. Balliere, Tindall, and Cox, London.
Uretsky, N. J. and Iversen, L.L. 1969. Effects of 6-hydroxydopa-
mine on noradrenaline-containing neurons in the rat brain.
Nature, 221, 557-559.
Wende, C.V. and Spoerlein, M.T. 1963. Oxidation of dopamine to
melanin by an enzyme of rat brain. Life Sci., 6, 386-392.
Wise, C.D., Berger, B.D., and Stein, L. 1970. Serotonin: A
possible mediator of behavioral suppression induced by anxiety.
Dis. Nerv. Sys., GWAN Supp. 31, 34-37.
Wise, C.D., Berger, B.D. and Stein, L. α-Noradrenergic receptors
for reward and serotonergic receptors for punishment in the rat
brain. Biol. Psychiat., in press, a.
Wise, C.D., Berger, B.D. and Stein, L. Benzodiazepines: Antianxiety activity by reduction of serotonin turnover in the brain.
Science, in press, b.
Wise, C.D. and Stein, L. 1969. Facilitation of brain self-
stimulation by central administration of norepinephrine. Science,
163, 299-301.
Wise, C.D. and Stein, L. 1970. Increased biosynthesis and
utilization of norepinephrine during self-stimulation of the
brain. Fed. Proc., 29, 485.
Wishner, J. 1965. Efficiency in schizophrenia. Bulletin de
l'association Inter. de Psychologie Appliquée, 14, 30-46.

SOME MACROMOLECULAR MECHANISMS IN CNS NEUROTRANSMITTER PHARMACOLOGY

AND THEIR PSYCHOBIOLOGICAL ORGANIZATION

Arnold J. Mandell, David S. Segal, Ronald T. Kuczenski

and Suzanne Knapp

Department of Psychiatry, University of California at

San Diego, La Jolla, California 92037

INTRODUCTION

Although it is often uncomfortable for us to do so, it is both salutary and liberating to acknowledge that in all science in general and in new science in particular, we are theoretically bound by the practical limits of our methodologies. When a limited methodology produces an apparent set of systematic observations and these data are equated with words implying function that have intuitive meaning to the scientist, we are on our way to a cosmology. For example, if inhibition of macromolecular biosynthesis in the brain can be associated with a specifiable defect in a task requiring delayed performance, we begin to talk about the chemistry of memory or the consolidation of experience (Barondes, 1970). If it appears that psychotropic drugs alter the release, re-uptake, or accessability of the receptor to biogenic amine putative transmitters (as defined by the myriad of "turnover technologies", Costa and Neff, 1970), we begin to develop theories relating amine dynamics to behavior and psychopathology. On one end, this system of tautologies is tied down by a series of assumptions worked out in vitro using drug interactions with synaptosomes and their biogenic amine uptake mechanisms as well as studies of peripheral sympathetic nerve endings treated with drugs. These findings are applied to the CNS via various measures of the "turnover" of the biogenic amine in question using disappearance rates, rates of conversion of substrate to product, or the pattern of metabolites before and after drug treatment (Costa and Neff, 1970). On the other side of the equation may be the drug induced or ameliorated psychopathological state in man, the drug altered behavioral state in animals, or in the case of model

105

peripheral systems, the function of an autonomically innervated
peripheral organ (Schildkraut, 1970).

The operations of the concomitant neuropharmacological theory
have as their primary mechanisms the impaired storage of amines
(reserpine), the impaired uptake of amines (tricyclic antidepres-
sants), the facilitated release of amines (cocaine), the impaired
amine-receptor interaction (phenothiazines), and the impaired
degradation of amines (monoamine oxidase inhibitors). As an added
note of complexity, a change in the rate of biosynthesis of amines
could result secondarily from the amount of amine available in the
presynaptic nerve ending for product-feedback inhibition of critical
biosynthetic enzymes. Within the limits of the "turnover technologies" most of the drugs of known CNS effect can be classed as
functioning via one or more of these mechanisms.

Our work over the past four years has led in the direction of
another set of alterable synaptic mechanisms involving the biogenic
amine neurotransmitters. These changes generally involve the
neurotransmitter biosynthetic enzymes. They occur following acute
or chronic alterations in the brain's environment (via chemical,
genetic, environmental, or physiological manipulation), and may or
may not be in the direction consonant with the findings of the
turnover studies. It has been a consistent temptation to relate
studies from the turnover era to the results of our studies of
neurotransmitter related macromolecular changes. However, since the
biological data collected is of very different sort and the experimental conditions very different, we have thus far resisted attempting to construct a close correlative relationship between these
measures. Since, in general, our measures relate to the macromolecular enzymatic potential of a system, we cannot comment about their
actual function in any given situation. Kopin and others (1968)
have noted that particularly in active peripheral systems, new synthesis of neurotransmitter is the most significant contribution to
the apparent functional neurotransmitter pool. If substrate were
not limiting (a situation that is difficult or impossible to establish for the specifically relevant subcellular substrate pool), the
amount or activity of the rate-limiting biosynthetic enzyme may be
the most meaningful measure of synaptic function. These and other
measures of macromolecular function (e.g., receptor sensitivity,
substrate uptake mechanisms, location and physical state of the
neurotransmitter biosynthetic enzymes) can only be related to studies
of the measures of neurotransmitter dynamics from other experiments
in a speculative way. In order to comment on the relationship between changes we have observed and other indicators of neurotransmitter dynamics, simultaneous measurements must be made. That such
relationships are not simple is revealed by the recent studies by
Dairman and Udenfriend (1970) and Weiner (1970) who have found that
measures of turnover may or may not relate to neurotransmitter
biosynthetic enzyme activity--differing between treatments or for

the same treatment at varying times. One could argue that measures
of macromolecular biosynthetic potential are artificial and remote;
that turnover measures in brain system during a state of experimental interest would be a more direct measure. From another point
of view, the multiple pools of substrate and neurotransmitter products, the effects of the experimental measures themselves (e.g.,
intracerebral injection under light anesthesia), and the multiple
potential artifacts of measures of turnover (Costa and Neff, 1970)
make the so-called "direct measures" less than direct. It is perhaps most realistic to view the set of measurements we shall report
here as an independent group of measures that may or may not relate
to the findings of the turnover studies. These studies may serve
as a source of converging data in the relatively new field of central
neurotransmitter pharmacology which may require other systems of
measures (single unit recordings, regional turnover studies) in
order to integrate into a whole picture.

The results of our current program of research will be presented
in the following order:

I.   Acutely Induced Changes in Neurotransmitter-Related
      Macromolecular Mechanisms
II.   Responses of Neurotransmitter-Related Macromolecular
      Mechanisms to Chronic Pharmacological and Nonpharma-
      cological Treatment
III. Some Thoughts Concerning the Integrative Organization
      of These Adaptive Changes
IV.   Some Implications for a Pathophysiology of Mental Disease

The work reported here will focus on tyrosine hydroxylase, tryptophan hydroxylase, and choline acetyltransferase, critical enzymes in
the biosynthesis of dopamine and norepinephrine, serotonin and acetylcholine, respectively. In addition, studies of receptor sensitivity
involving intraventricular norepinephrine induced behavioral activation and brain adenylcyclase, as well as studies of synaptosomal uptake of a neurotransmitter precursor, tryptophan will be reported.
The methods used for drug administration, environmental and genetic
manipulations, enzyme studies, behavioral monitoring and amino acid
uptake studies have been reported elsewhere and are referred to in
the appropriate sections.

I. ACUTELY INDUCED CHANGES IN NEUROTRANSMITTER-RELATED
MACROMOLECULAR MECHANISMS

A. Drug Induced Alteration in Physical State of a
Neurotransmitter Biosynthetic Enzyme

The diversity of reports concerning the physical state of
tyrosine hydroxylase appears related to the organ or region involved.

Whereas is appears that adrenal tyrosine hydroxylase is soluble in its native form (Wurzburger and Musacchio, 1971) and only particulate as a function of its conditions of preparation, the brain enzyme appears to be in both a soluble and particulate native state. McGeer et al. (1965) were the first to report a synaptosomal fraction from striate cortex to be high in tyrosine hydroxylase. Fahn et al. (1969) then reported that bovine caudate nucleus tyrosine hydroxylase was associated with synaptic vesicles, implying that some sort of entrapment in the vesicles accounted for the particulate nature of brain tyrosine hydroxylase. We (Kuczenski and Mandell, 1972; Mandell et al., 1972, in press) have recently shown that most of the particulate tyrosine hydroxylase from rat striate cortex in hypotonically shocked synaptosomes and synaptic vesicles sediments with the synaptosomal membranes in discontinuous sucrose density gradients. In addition, treatment which almost completely solubilizes synaptosomal DOPA decarboxylase and destroys intact synaptosomes as observed in electron micrographs does not alter the membrane-bound character of a siginficant fraction of the striate tyrosine hydroxylase. Areas with known high density of catecholamine cell bodies such as the midbrain or locus ceruleus have very little particulate tyrosine hydroxylase. The particulate fraction of brain tyrosine hydroxylase appears related to binding to synaptosomal membranes, not synaptosomal entrapment, and is specific to brain regions dense in catecholamine nerve endings.

In light of the stable physical character of striate particulate tyrosine hydroxylase, it was significant that methadrine produced a shift in the subcellular distribution of this enzyme. After methadrine administration, a significant amount of enzyme activity appeared to shift from the 11,000 x G supernatant to the synaptosomal fractions with no change in total measurable activity (Mandell et al, 1972). Table I summarizes the results of a typical experiment in which the caudate areas were studied two hours following the subcutaneous administration of methamphetamine hydrochloride, 1 mg/kg. Note that there is a nonsignificant change in total activity at the same time that there is a shift of the measurable activity into the P_2B and P_2C fractions. Because this alteration occurred to a similar extent in both these fractions and the P_2A fraction manifested no enzyme activity, further studies were carried out on the whole crude mitochondrial pellet. Figure 1 represents the results of a series of experiments using this cruder separation. Following a dose of 5 mg/kg of methamphetamine, a shift of tyrosine hydroxylase from the soluble to the particulate fraction was observed which was maximal at 30 minutes and lasted six hours. There was no significant change in total measurable activity at any time. Midbrain tyrosine hydroxylase (which, using the same fractionation methods appeared to be about 80% soluble and 20% synaptosomal) did not manifest this shift with drug treatment although it is possible that a comparable relative shift to the synaptosomal fraction would be immersed in the

**THE EFFECT OF AMPHETAMINE ON THE SUBCELLULAR
DISTRIBUTION OF TYROSINE HYDROXYLASE***

	% OF TOTAL ACTIVITY	
	SALINE	AMPHETAMINE
11,000 x G SUPERNATENT	69	36
P_2A	0	0
P_2B	16	35
P_2C	14	27

	TOTAL ACTIVITY $\mu\mu M$ DOPA/mg prot/hour
SALINE	102.3 ± 19.4
AMPHETAMINE	81.0 ± 16.7

*Four hours after 15 mg/Kg; means of
three pools of three caudates.

Table I.   The total activity and subcellular distribution of rat
striatal tyrosine hydroxylase studied two hours after the administration of methamphetamine hydrochloride, 1 mg/kg subcutaneously or
an equivolume amount of saline.   The values equal the means of three
pools of three caudate areas each.   The subcellular distribution is
expressed as the mean percent of total measurable activity, contained
in each fraction.   Amphetamine produced no effect on total activity
which is expressed as $\mu\mu$ moles DOPA synthesized per hour.   Note the
shift of enzyme activity from the soluble to the "synaptosomal" and
"mitochondrial" fractions.

error term in the midbrain.   Chronic treatment with amphetamine with
increasing daily doses (5 mg, twice a day to 50 mg twice a day)
resulted in the disappearance of the amphetamine induced shift on
the eighth day.   The time of the disappearance in this shift corre-
sponded with the development of tachyphylaxis to the behavioral
effects of the drug (Mandell et al., 1971, 1972).   The shift was not
observed when methamphetamine was added before homogenization in
vitro.   This change was dose-related reaching a maximum effect at
5 mg/kg.   At high doses, the shortest latency observed for the shift
was about 10 minutes after drug administration and the longest duration observed was about eight hours.   Imipramine, footshock, electroconvulsive shock and intraventricular infusion of norepinephrine
did not produce this effect whereas the administration of α-methyltyrosine did.

A similar short-latency shift in enzyme activity has been
observed for other drugs but in these cases, there is an increase
in total measurable enzyme activity, all in the particulate frac-

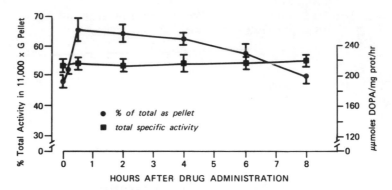

Figure 1. The total measurable striatal tyrosine hydroxylase activity
(μμmoles DOPA synthesized per hour) and the percent of that total in
the post low-speed spin, 11,000 x G pellet at various times after
methamphetamine hydrochloride, 5 mg/kg. Each point represents the
mean and the standard error of the mean of six caudate pairs (six
animals). Whereas total measurable activity was not changed by the
administration of methamphetamine, the percent in the pellet rose
rapidly reaching its peak in 30 minutes and remained significantly
different from saline controls (which were not affected) until six
hours.

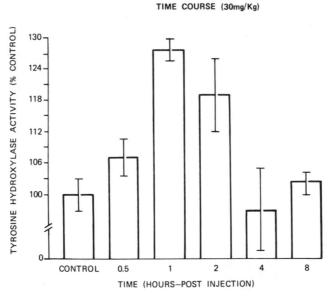

Figure 2. Effect of a single dose (ip) of propranolol, 30 mg/kg, on
total striatal ("caudate") tyrosine hydroxylase activity at various
times after drug administration. Highest values were obtained at
the one hour interval with return to control values at four hours.
Both the one and two hour interval determinations were significantly
greater than control ($p < .05$); n = 6 for each group.

tions. Figure is from data reported more extensively by Sullivan
et al. (1972) demonstrating an acutely induced increase in striate
tyrosine hydroxylase activity following the administration of
propranolol (30 mg/kg). A similar kind of change has been observed
following the acute administration of large doses of reserpine (5
mg/kg) which appeared to last considerably longer (at least 24 hours).
In these experiments, there did not seem to be a significant associated decrease in the activity of the striate soluble fraction.

It appears that for methamphetamine, reserpine and propranolol
an acute change in subcellular distribution may be induced by drug
treatment. The effects of methamphetamine and propranolol were not
inhibited by pretreatment with sufficient cycloheximide (2 mg/kg) to
inhibit 80% of the incorporation of ^{14}C-leucine into the brain tri-
chloroacetic acid precipitable protein. These changes may be due to
effects on the enzyme-synaptosomal membrane interaction via drug-
induced changes in tissue ions. For example, we (Kuczenski and
Mandell, 1972a, 1972b) have recently shown that when free Calcium
(Ca^{++}) is present during the homogenization and fractionation procedures used to isolate striate tyrosine hydroxylase, there is a
marked increase in synaptosomal membrane bound enzyme. The kinetic
and regulatory ramifications of these binding studies will be discussed in the following section. It should be noted that divalent
cations have been directly or indirectly implicated in the actions
of propranolol and reserpine (Boyaner and Radouco-Thomas, 1971).
For the amphetamines, it is tempting to relate the drug-induced
shift in subcellular distribution of striate tyrosine hydroxylase
and its subsequent tachyphylaxis to the primary stimulatory effects
of the drug. Whether these changes are truly primary or are in
effect a fast-acting mechanism of adaptation to another drug-induced
effect (such as a drug-induced alteration in the re-uptake or binding
of a biogenic amine in the presynaptic nerve ending) awaits further
work.

B. Regulatory Properties Related to the Physical State of
Neurotransmitter Biosynthetic Enzymes

As noted above, rat brain tyrosine hydroxylase appears to exist
in two distinct physical forms, a soluble and a membrane bound particulate form. These two forms appear related to specific brain
regions, with the particulate form present in areas high in catecholamine nerve endings and the soluble predominantly in areas dense
in catecholamine cell bodies. We (Kuczenski and Mandell, 1972b) have
recently shown that soluble brain tyrosine hydroxylase was markedly
activated by low concentrations of a specific sulfated mucopolysaccharide, heparin, but not by the chondroitin sulfates nor by hyalu-
ronic acid. Figure 3 demonstrates that heparin decreases the Km for
the synthetic cofactor of the soluble enzyme as well as increases the
apparent Vmax. On the other hand, heparin has no effect on the
membrane bound form. The stimulation of the membrane binding by a

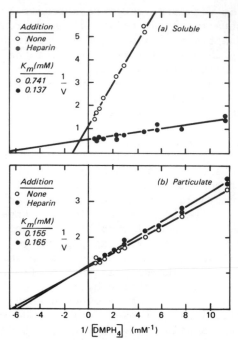

Figure 3.   Lineweaver–Burke plot of the activity of caudate (A) soluble tyrosine hydroxylase and (B) particulate tyrosine hydroxy-lase as a function of $DMPH_4$ concentration in the presence and absence of 0.0286 mg/ml heparin.   The assay mixture contained 3×10^{-6} M tyrosine-3,5-H^3 (containing 10^6 cpm). 1.1×10^{-3}H $DMPH_4$, 0.014 M β–mercaptoethanol and 8.76×10^{-6} M $FeSO_4$.   Each assay contained 250 μgm protein.

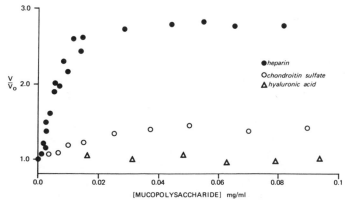

Figure 4.   Effect of polysaccharides on the activity of tyrosine hydroxylase.   See legend for Figure 3 for assay conditions.

specific sulfated mucopolysaccharide suggests that such a site may
be present in nerve ending membranes. Vos et al. (1968) have
reported sulfated mucopolysaccharides in synaptosomal membranes.
The relationship between membrane bound state and increased affinity
for cofactor was also shown (Kuczenski and Mandell, 1972a) to be
true for the product-feedback inhibitors, dopamine (DA) and norepinephrine (NE), that compete for the $DMPH_4$-cofactor site. Dopamine
has a higher affinity than norepinephrine for both the soluble and
the particulate forms of the enzyme. The activation by heparin of
the soluble enzyme (Figure 4) exhibits sigmoidal (cooperative) binding with a Hill slope (n_H) near 2. The Ka for heparin was 0.5 µM
with a molecular weight of 16,000.

Heparin activation closely simulated membrane binding as
measured by the resemblance of the pH curve of the heparin activated
enzyme to the membrane bound but not the soluble enzyme, the similarity of the Km for $DMPH_4$ of the heparin activated and membrane
bound enzyme, and the similarites of these two effects suggest that
both forms of activation may proceed by a common mechanism. In
addition, we have recently succeeded in completely solubilizing
particulate tyrosine hydroxylase. Complete recovery of activity can
be observed only by addition to the solubilized enzyme of heparin.
Further, preliminary data resulting from in vitro produced shifts of
enzyme from the soluble to the particulate fraction using divalent
cations (Ca^{++}, Mg^{++}), reveal an increase in total activity. It thus
seems that the drug-induced shift in the subcellular distribution of
striate tyrosine hydroxylase from the soluble to the particulate
fraction as summarized in Section I-A could be associated with
membrane-binding activation of the enzyme. Short latency increases
in nerve ending enzyme activity may result from such physical state
changes and their kinetic ramifications.

On the other hand, in vitro studies of striatal tyrosine hydroxylase have from time to time, depending upon the conditions for
subcellular fractionation and assay, revealed what appears to be an
occlusive binding to membranes. In addition, recent work in our
laboratory has revealed that the monomeric form (M.W. 49,000) of
tyrosine hydroxylase (native M.W. 200,000) could be inhibited by
heparin which manifested the same specificity of binding when compared with the chondroitin sulfates (Kuczenski and Mandell, 1972).
It is tempting to speculate that some kinds of membrane binding may
alter the conformation of the enzyme so as to partially mask an
active site. It seems possible that alterations in the physical
state of brain tyrosine hydroxylase may either serve to activate or
inhibit the enzyme.

C. A Drug-Sensitive Uptake Mechanism for the Substrate of
   a Neurotransmitter Biosynthetic Enzyme

The early studies of tryptophan-5-hydroxylase in mammalian brain

made use of particulate enzyme from the "crude mitochondrial pellet"
prepared from homogenization in 0.32 M sucrose (Grahame-Smith, 1967)
or soluble enzyme prepared from hypotonic homogenization followed
by 12,000 x G or 50,000 x G centrifugation (Ichiyama et al., 1970).
Although the studies by Grahame-Smith and Ichiyama et al. have
acknowledged the presence and some of the characteristics of both
soluble and particulate tryptophan-5-hydroxylase, these two enzyme
forms have not been systematically compared as to their kinetics,
regional distribution, or sensitivity to drugs. The most common
area of the brain used to study this enzyme has been the midbrain
which contains a high concentration of serotonergic cell bodies in
the medial raphe as demonstrated by Dahlstrom and Fuxe (1964). Here
the relatively small size of the particulate fraction has led workers
to ignore this physical form of the enzyme. Our studies of trypto-
phan-5-hydroxylase activity in other brain regions with and without
drugs have led us to conclude that the particulate form of the
enzyme constitutes a significant fraction of the total activity in
areas in which serotonergic nerve endings are prominent (such as
the frontal cortex, septal area, caudate and the lumbosacral cord)
and the soluble form of the enzyme is predominant in midbrain and
pons-medulla area where serotonergic cell bodies have been reported
(Dahlstrom and Fuxe, 1964). Since these two physical forms of the
enzyme manifest different characteristics in vitro as well as different responsivity to drugs in vivo, we began the in vitro characterization of both physical forms and parallel studies of their responsiveness to various agents.

The importance of substrate supply in the regulation of serotonin
biosynthesis has been emphasized due to the disparity between the Km
of the tryptophan-5-hydroxylase (5×10^{-4} M) for substrate and the
reported cytoplasmic concentration of tryptophan in brain, 4×10^{-5}
M as reported by Peters et al. (1968). Recent studies by Fernstrom
and Wurtman (1971) and Grahame-Smith (1971) have demonstrated a dose
related relationship between administered tryptophan, plasma and brain
tryptophan levels and serotonin brain levels within a relatively
wide range of values.

Our work has focused on the simultaneous determination of drug
and environmental effects on both midbrain (soluble) and septal
(particulate) enzyme activity as well as effects on the in vitro
uptake of radioactive tryptophan. We have found the latter to be
an energy dependent, saturable process with a Km in the range of
5×10^{-5} for tryptophan. It was sensitive to drug effects. Details
of these techniques and findings are either in press or in preparation (Knapp and Mandell, 1972 in press, 1972 in preparation). The
following material focuses on drug effects on the nerve ending
serotonin biosynthetic process as influenced by drug effects on the
nerve ending uptake process.

Preparation of the synaptosomal tryptophan hydroxylase required careful homogenization in 0.32 M sucrose for the maintenance of maximal activity. Homogenization was followed by appropriate centrifugation to prepare the crude mitochondrial pellet. This pellet was suspended in 0.32 M sucrose and used as the source of the particulate enzyme. The "particulate" fraction, in addition to being complete (requiring no added cofactors) for the enzymatic conversion of tryptophan to serotonin, manifested an active uptake process as reported by Grahame-Smith and Parfitt (1970) with a far greater affinity for the tryptophan substrate than that reported by that group. This was probably due to our use of a region with a known relative high density of serotonin nerve endings rather than a whole brain synaptosome preparation. The continuity between the cell body and nerve ending enzyme will be discussed in a subsequent section. The following are a series of studies of the effects of morphine and cocaine on these two forms of tryptophan hydroxylase that demonstrated the acute effects of drug on the substrate-uptake process.

Considerable attention has been focused on the relationship of narcotic drug administration and alterations in serotonin turnover. This has been measured many ways; for example, the rate of the accumulation of serotonin after the administration of a monoamine oxidase inhibitor (Tozer et al., 1966). Alteration in serotonin turnover has been related by some authors to be the addictive process itself in that it was reported by Way et al. (1968) that the administration of PCPA prevented some manifestations of withdrawal in mice, which was not confirmed by Cheney et al. (1971). We have seen an apparent differential response to acute and chronic morphinization of rats when effects on midbrain soluble and septal particulate enzyme activity were compared. Figure 5 represents a summary of such experiments. In the acute experiments, morphine sulphate, 10 mg/kg was administered subcutaneously and the animals were sacrificed three hours later. In the chronic experiments, a pellet containing 75 mg of morphine sulfate was implanted subcutaneously for five days. Twelve animals were studied in each group. Note that the acute effects of morphine administration resulted in a significant decrease in the measurable activity of septal particulate but not the midbrain soluble enzyme. The chronic administration of morphine by pellet implantation resulted in a significant increase in apparent septal particulate enzyme activity but no change in the soluble enzyme. Figure 6 demonstrated that morphine sulfate at concentrations of 5 x 10^{-6} to 5 x 10^{-4} increasingly inhibited the soluble enzyme in vitro. Particulate enzyme activity is likewise shown to be significantly reduced. Morphine in this range of concentrations failed to alter the uptake of 3-^{14}C labelled tryptophan. Scrafani et al. (1969) and Navon and Lajtha (1970) recently reported a narcotic drug concentration mechanism in synaptosomes, inhibited by narcotic blockers that may account for the difference in the acute effect of morphine administration on the soluble and particulate enzyme as in Figure 5.

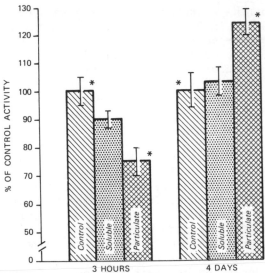

Figure 5.   The effect of a single dose of morphine sulfate (10 mg/kg) and a pellet implant for four days on midbrain soluble and septal particulate tryptophan hydroxylase.   N = 12 per group.   An asterisk indicates a statistically significant change from controls with a p < .05.   See text.

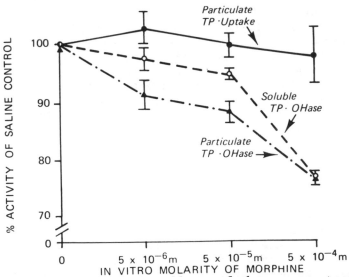

Figure 6.   A comparison of the effects of three concentrations of morphine sulfate on the hydroxylation and uptake of tryptophan by synaptosomes and the hydroxylation of tryptophan by midbrain soluble enzyme in vitro.   Each point represents the mean ± S.E.M. for 3 experiments.   Note that whereas morphine does not affect the uptake process, it does inhibit hydroxylation in the soluble and even more in the particulate where morphine is probably concentrated.   See text.

The effects of the acute administration of cocaine supplies a contrasting example of a differential drug effect on uptake of substrate and hydroxylation by serotonin nerve endings. Figure 7 summarizes experiments in which cocaine, 30 mg/kg was administered subcutaneously one hour prior to sacrifice. Note that whereas there is no measurable effect on midbrain soluble enzyme activity, there was a small but significant decrease in the apparent activity of septal particulate enzyme. In subsequent efforts to discriminate drug effects on tryptophan uptake from hydroxylation, contrasting results from those seen with morphine were observed. Figure 8 shows that cocaine at concentrations of from 1×10^{-5} M to 1×10^{-3} M did not affect the activity of the soluble enzyme whereas it reduced significantly the apparent activity of septal particulate tryptophan hydroxylase. Similar concentrations of cocaine also inhibited the uptake of $3-^{14}$C tryptophan into synaptosomes. Thus it appears that whereas morphine reduces measurable nerve ending tryptophan hydroxylase activity by inhibition of enzymatic hydroxylation, cocaine produces the same net effect by reduction of the substrate available to the enzyme in nerve endings by inhibiting uptake.

Recent work in our laboratories has demonstrated that α-methyl-tryptophan, leucine, and phenylalanine are more potent inhibitors of the tryptophan uptake process than DOPA. In addition, some preliminary evidence has suggested that pretreatment with a tryptophan load 2-4 hours before sacrifice leads to an apparent increase in synaptosomal serotonin biosynthetic activity seen to be secondary to changes induced by the load in the uptake process. It is clear from these and other studies that the amino acid uptake process into synaptosomes and drug and environmental effects on it are potentially a significant source of acute changes in neurotransmitter-related macromolecular function in brain.

D. A Drug Initiated Sequence of Changes in a Neurotranmitter Biosynthetic Enzyme - A Model Time Base for Macromolecular Adaptation

Experiments to elucidate the effects of parachlorophenylalanine (PCPA) on serotonin biosynthesis has revealed a series of drug interactions with serotonin biosynthetic enzymes that suggest that both subcellular region and time after drug administration are important dimensions in the understanding of the mechanisms underlying the acute effects of drugs. The mechanisms of action on the serotonin biosynthetic process appear quite complex. Thus far, it appears that two general mechanisms can account for the ability of PCPA administration to reduce brain serotonin. In vitro, PCPA has been shown to competitively inhibit midbrain enzyme with regard to substrate (Jequier et al., 1967). There have been reports that, depending upon the dose and time after administration, it is both a reversible competitive inhibition for substrate and an irreversible

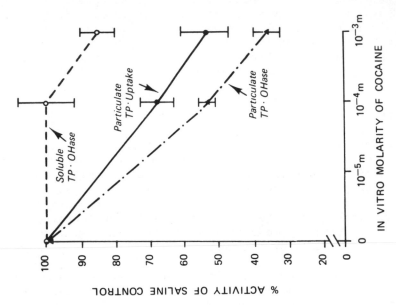

Figure 8. A comparison of the effects of three concentrations of cocaine hydrochloride on the hydroxylation and uptake of tryptophan by particulate and hydroxylation by soluble enzyme. Each point represents the mean ± S.E.M. for six experiments. Note that cocaine does not inhibit soluble enzyme significantly; it does inhibit the apparent particulate uptake process and therefore the apparent enzymatic hydroxylation by the particulate enzyme as well. See text.

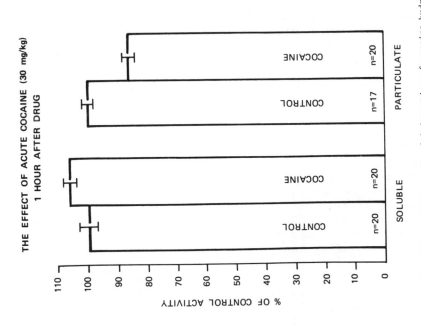

Figure 7. The effects of the acute administration of cocaine hydrochloride (30 mg/kg) on soluble and particulate tryptophan hydroxylase expressed as percent of the control levels. Note that there is a small but significant decrease in apparent particulate activity ($p < .05$).

non-competitive inhibitor (Gal et al., 1970). Gal et al. (1970) using liver phenylalanine hydroxylase as a model has suggested that the undialyzable inhibition was due to the incorporation of this abnormal amino acid into the enzyme protein near the active site. A recent series of experiments conducted in our laboratory (Knapp and Mandell, 1972) appeared to define a sequence of PCPA effects and present a model time base for a sequence of changes in the serotonin cell bodies and their nerve endings.

PCPA (300 mg/kg) was administered to Sprague-Dawley rats (180-200 gm) and groups of 12 rats each were sacrificed at 30 minutes, 1 hour, 4 hours, 2 days, 8 days, 13 days, and 18 days following drug administration. Tryptophan hydroxylase activity was determined in the midbrain-soluble fraction (80% of the total midbrain activity) and septal-particulate (100% of the total septal activity) fraction. The soluble enzyme activity was determined with and without dialysis. The particulate enzyme was not dialyzed due to its instability. Figure 9 is a summary of the findings expressed as per cent of control values. At 30 minutes and 1 hour, there was a marked decrease in midbrain tryptophan hydroxylase activity which returned to control levels following dialysis (and is indicated in Figure 9 as within the control range). The decrease of midbrain activity at four hours was only partially reversible by dialysis and reached a maximum for these groups of measurements at two days. The values at this time were not altered by dialysis. A gradual return to control levels in the midbrain was noted between 8 and 13 days after the administration of the drug. The septal synaptosomal enzyme demonstrated a noticeable decrease by 30 minutes, reached a maximum by one hour, and returned to control levels by four hours. This early inhibition of septal tryptophan hydroxylase activity was shown to be due to the competitive inhibition of tryptophan substrate uptake into serotonin nerve endings for which PCPA has a Ki of 9.8×10^{-6} M. The maintenance of activity in the control range in the septal region was noted at 2 and 8 days followed by a measurable decrease in activity at 13 and 18 days after drug administration. What appears in this series of studies is a sequence of four mechanisms by which PCPA inhibits serotonin biosynthesis: First, a competitive inhibition of the uptake of tryptophan into the serotonin nerve ending; second, a competitive inhibition of midbrain cell body enzyme; third, an irreversible inhibition of tryptophan hydroxylase enzyme in the cell body (the incorporation of an abnormal amino acid into the enzyme by the protein synthetic machinery; and fourth, the axoplasmic flow of this defective enzyme into the serotonin nerve endings at the rate of 1-2 mm/day (a total distance of about 1.5 cm). The latency from the time of drug administration to time of arrival of the drug altered enzyme at the nerve endings may bear some relationship to the latency of effects of the adaptive mechanisms induced by chronic pharmacological treatment (as discussed in the following section).

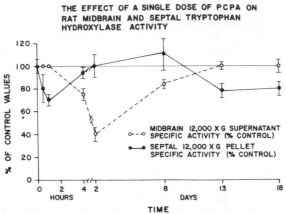

Figure 9.   The activity of midbrain and septal tryptophan hydroxylase following the acute administration of PCPA, 300 mg/kg.   Note
the initial reversible decrease in septal enzyme activity followed
by the return to control levels and then a delayed fall.   At the
same time, midbrain enzyme is decreased more slowly and more pro-
foundly with a delayed return to control levels.   These data are
seen as exemplifying an acute effect on substrate availability in
particulate enzyme followed by an apparent axoplasmic flow from midbrain to septum of defective tryptophan hydroxylase.   See text.

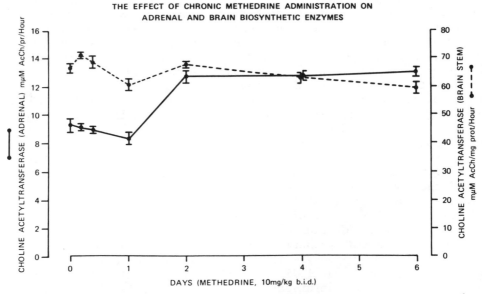

Figure 10.   The activity of choline acetyltransferase per adrenal
pair at 0, 6, 12, 24, 48, 96, and 144 hours after the initiation of
twice-a-day administration of methadrine, 10 mg/kg.   Each point
represents the mean, ± the standard error of the mean for each
group of six animals.   See text.

The probability that the neurotransmitter biosynthetic enzymes travel from the cell body to nerve endings by slow flow was also suggested by our studies of the drug effects on choline acetyltransferase. Figure 10 demonstrates that whereas amphetamine administered to White Leghorn Chicks results in the elevation of brain stem and cervical cord choline acetyltransferase activity within six hours, the "wave" of increased enzyme activity does not reach the preganglionic cholinergic nerve endings to the adrenal medulla for over 48 hours. The distance from the cord to the paravertebral area where the adrenal lies in the chick is less than 0.5 cm.

It is possible that the slow flow rate of neurotransmitter biosynthetic enzymes creates an unalterable delay between the primary effects of drugs on macromolecular mechanisms at the synapse and the secondary changes involving alterations in the rate of synthesis or degradation of neurotransmitter biosynthetic enzymes and their transport to their locales of major function, the nerve endings. This would, of course, be dependent on the length of the relevant cells and the character of the pathway (e.g., mono or polysynaptic) as well as the responsivity of the system. As of the present writing, there has been no evidence adduced in our laboratory or others that the the rate of this flow can be altered by pharmacological or other treatments. This sort of sequence should be kept in mind in the following discussion of the responses of neurotransmitter related macromolecular mechanisms to chronic pharmacological and nonpharmacological treatment.

II. RESPONSES OF NEUROTRANSMITTER-RELATED MACROMOLECULAR MECHANISMS TO CHRONIC PHARMACOLOGICAL AND NONPHARMACOLOGICAL TREATMENT

A. Regulation of Catecholamine Biosynthesis in the Sympathetic Nervous System

In addition to the rapid adjustment of the rate of catecholamine biosynthesis which is mediated by the feedback inhibition of tyrosine hydroxylase by NE, recent evidence also indicates the existence of another, slower regulatory process which involves alterations in the level of tyrosine hydroxylase activity. Support for this latter process stems primarily from the elegant series of studies done By Thoenen, Mueller and Axelrod (Mueller et al., 1969a, 1969b) in which drugs that impaired sympathetic transmission (such as 6-hydroxydopamine, reserpine and phenoxybenzamine) produced an increase in the activity of tyrosine hydroxylase in sympathetic ganglia and/or in the adrenal gland. This elevation of enzyme activity appeared to be due to an increased rate of synthesis of new enzyme. This effect was mediated by the prolonged reflexive increase in sympathetic nerve activity since it could be prevented by cycloheximide or decentralization. In addition, Dairman and Udenfriend

(1970) have shown that the increased tyrosine hydroxylase activity (as measured in tissue homogenates) correlated with an in vivo acceleration of catecholamine synthesis as measured by the conversion of isotopic tyrosine to norepinephrine. They found that phenoxybenzamine produced these effects. A ganglionic blocking agent prevented the enhanced conversion of tyrosine to catecholamines after one day of phenoxybenzamine treatment, but had a comparably smaller effect following more prolonged treatment with phenoxybenzamine. These results were interpreted as indicating that the accelerated accumulation of labeled catecholamines observed after one day was due primarily to increased nervous activity and that the pentolinium-resistant increase was due to an elevation in the concentration and/or activity of the enzyme tyrosine hydroxylase.

If these increases in tyrosine hydroxylase activity represented a compensatory or homeostatic mechanism in response to an impairment in the state of adrenergic transmission, then it might be expected that long-term facilitation of transmission ought to result in a decline in enzyme activity. In support of this contention, Dairman and Udenfriend (1970) have shown that tyrosine hydroxylase activity was markedly decreased in the rat adrenal and arterial walls following the repeated administration of L-dopa. This alteration in enzyme activity may be mediated by an L-dopa induced decline in sympathetic nerve activity (Whitsett et al., 1970). Also consistent with this interpretation is the finding that genetically hypertensive rats have both a low level of both blood vessel tyrosine hydroxylase (Tarver et al., 1971) and sympathetic nerve activity (Louis et al., 1969) when compared to normotensive controls.

Similar increases and decreases in tyrosine hydroxylase activity have been reported following nonpharmacological treatment. Axelrod and his coworkers (1970) have reported that mice deprived of psychosocial stimulation for six months showed a significant decrease in adrenal tyrosine hydroxylase while stimulated mice had elevated enzyme levels. In addition, prolonged immobilization stress (Kvetnansky et al., 1970) or cold-exposure of rats (Thoenen, 1970) led to increased tyrosine hydroxylase activity in the adrenal medulla as well as the superior cervical ganglia. The elevation in enzyme activity could be prevented by transecting the preganglionic nerves to the adrenal and sympathetic ganglia. Similar changes have recently been reported for dopamine-β-hydroxylase (Molinoff et al., 1970). The studies of the peripheral sympathetic nervous system have suggested that sustained changes in adrenergic activity may result in the modulation of the amount or activity of neurotransmitter biosynthetic enzymes.

Over the past several years we have examined the effects of various chronic treatments on brain tyrosine hydroxylase, tryptophan hydroxylase, and choline acetyltransferase activity. The results

of these studies, reviewed in the following section, are consistent
with the view that as in the periphery, prolonged alterations in
synaptic activity may result in alterations in the amount or activity of the relevant neurotransmitter biosynthetic enzymes. In
addition, the direction of these enzymatic changes appears generally
to be in the direction of a homeostatic compensation for the induced
alteration in synaptic activity. In general, the magnitude of
changes in enzyme activity observed in the brain are smaller than
seen in peripheral systems. Representative changes in brain enzymes
are in the 10-30% range whereas those in sympathetic nerves may be
over 50% and in the adrenal medulla over 100%.

> B. Chronic Pharmacological and Nonpharmacological Effects
>   On Brain Enzymes and Behavior

(1) <u>Treatments which initially impair transmission.</u>
        a. <u>Reserpine</u> is believed by many to produce behavioral depres-
sion by depleting the brain of catecholamines (CAs) (Schildkraut and
Kety, 1967). However, we and others have found that rats chronically
treated with reserpine in appropriate doses eventually exhibit marked
behavioral hyperactivity (Figure 11) and this hyperactivity occurred
at a time when the brain CA concentrations were significantly reduced.
This correlation between behavioral depression and low brain catecholamines was questioned by some due to other chemical indices suggesting a reserpine-induced increase in catecholamine turnover
(Mandell and Spooner, 1968). This issue, although still a lively
one, has been partially resolved by several lines of evidence indicating that only a small portion of the total brain CA level is
functionally significant. Furthermore, it appears that the functionally important CA pool is relatively small and is maintained primarily by <u>de novo</u> synthesis. Studies have shown that newly synthesized amines are preferentially released when adrenergic neurons are
activated (Kopin et al., 1968). This finding has placed new empha-
sis on the neurotranmitter biosynthetic process as critical in the
regulation of synaptic neurotransmitter levels and particularly
the level of the amount or activity of the enzymes. As in the periphery, it was speculated that decreases in adrenergic transmission
may be communicated back to the adrenergic neuron with resulting
alterations in tyrosine hydroxylase activity. This might even produce a transient excess (over compensation) of synaptic amines while
the total level of brain catecholamines were still reduced. This
was the premise entertained to explain the behavioral hyperactivity
induced by the chronic administration of reserpine. Figure 12 shows
that, in fact, midbrain tyrosine hydroxylase activity was significantly increased in animals that had received reserpine (0.5 mg/kg,
ip) for 9 days when compared to appropriate controls. Similar substantial increases in striatal enzyme activity were also observed,
but delayed to some extent. This disparity in time course between
the two regions may reflect the time required for the axoplasmic

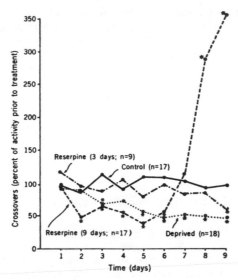

Figure 11.   The effects of four schedules of reserpine treatment
(see text) or food deprivation on the midbrain tyrosine hydroxylase
activity of rats.   Each bar represents the mean ± s.e.m. (brackets)
of the indicated number of observations.   Only eight to nine days
of chronic reserpine treatment induced a significant increase in
enzyme activity (p < 0.05, Mann–Whitney U test).

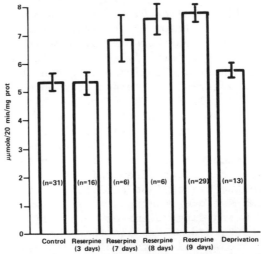

Figure 12.   The effects of four schedules of reserpine treatment
(see text) or food deprivation on the midbrain tyrosine hydroxylase
activity of rats.   Each bar represents the mean ± s.e.m. (brackets)
of the indicated number of observations.   Only eight to nine days
of chronic reserpine treatment induced a significant increase in
enzyme activity (p < 0.05, Mann–Whitney U test).

flow of newly synthesized enzyme from a cell body enriched area
(midbrain) to an area dense in catecholamine nerve terminals
(striatum). It has been shown that the weight loss associated with
chronic reserpine administration was not responsible for the changes
in either the behavior or the enzyme activity (Segal et al., 1971).

These results demonstrated a temporal correlation between the
prolonged depletion of brain CAs by chronic reserpine treatment and
subsequent enhancement of behavioral responsivity and tyrosine
hydroxylase activity in the midbrain and neostriatum. On the basis
of these findings we speculated that as in peripheral sympathetic
systems, changes in neurotransmitter biosynthetic enzymes may have
a compensatory or adaptive function related to the functional activity of their respective transmitter substances.

b. Thyroid State. In order to test the generality of the above
findings, other treatments were devised. One of these involved
the manipulation of the thyroid status in rats. In addition to
reports of thyroid induced alterations in brain norepinephrine
turnover in rats (Prange et al., 1969, 1970), its use as a poten-
tiator of tricyclic antidepressant action in man (Prange et al.,
1969, 1970) suggested this was a relevant dimension.

Drawing on work in the literature suggesting that thyroid
hormone sensitizes peripheral adrenergic receptors (Krishna et al.,
1968), it was hypothesized that thyrozine-treated rats would be
supersensitive to brain synaptic catecholamines and a compensatory
lowering of tyrosine hydroxylase would result. Thyroidectomy would
lead to an elevation in the enzyme. In order to test this hypothesis, we examined the spontaneous motor activity and midbrain tyrosine hydroxylase activity in rats injected with 0.4 mg/kg of sodium-
L-thyroxine (s.c.) daily for 10 days, or thyroidectomized three
weeks prior to the onset of the experiment. Control and thyroidectomized rats received daily saline injections. The results showed
that thyroidectomy produced a marked increase in midbrain tyrosine
hydroxylase activity when compared to control values (Figure 13).
The behavioral activity was reduced in the experimental animals
although this difference did not attain statistical significance.
A small group of thyroidectomized animals were shown to be hypo-
sensitive to intraventricularly infused norepinephrine. It is
therefore likely that the increase in the specific activity of the
rate-limiting enzyme for CA biosynthesis may be responsible for
producing the increase in the conversion of isotopic precursor to
NE reported previously (Prange et al., 1970). This was probably
triggered by a desensitization of the adrenergic receptors but
showed sufficient compensation to result in almost normal spontaneous
motor activity. Although thyroxine treatment resulted in both an
increase in motoric activity and increased sensitivity to intraventricularly administered norepinephrine, there was not the expected

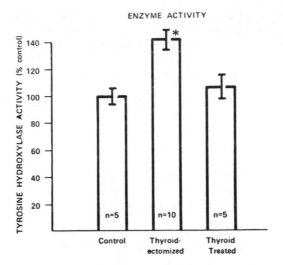

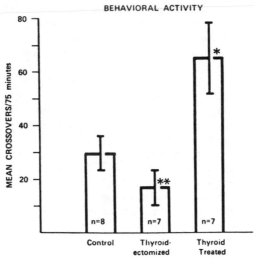

Figure 13.   The top effects of thyroid treatment and thyroidectomy
on the specific activity of midbrain tyrosine hydroxylase (expressed
as % control).   Thyroid treated animals received 0.5 mg/kg Na-L-
thyroxine for 10 days, and showed no change in enzyme activity.
Thyroidectomized animals were sacrificed 3 weeks after thyroidectomy
and showed a significant increase in enzyme activity ($p < 0.01$).
Brackets represent the standard error of the mean.   (Bottom) the
effects of thyroid treatment on free field behavioral activity.
Activity during a 75 min period was measured as crossovers from one
quadrant to another.   Thyroid treated rats showed a significant increase in activity (* = $p < 0.02$), and thyroidectomized rats showed
a decrease in activity, although not significant (** = $0.05 < p <
0.1$).   Brackets represent standard error of the mean.

compensatory decrease in midbrain tyrosine hydroxylase activity.
Some of the issues involved have been discussed elsewhere (Emlen,
1972). Very recent work with a range of compounds that potentiate
central adrenergic receptors has suggested that well over two or
three weeks of chronic treatment may be required to see decreases in
midbrain tyrosine hydroxylase. It is possible that the thyroxine
treated group was studied prematurely.

    c. Environmental Deprivation. In an attempt to use a more
physiological manipulation of the brain's biogenic amine systems
than a drug or a hormone, rats were isolated for 4, 8 or 16 days and
their regional brain tyrosine and tryptophan hydroxylase activities
were compared to animals housed in groups of six, under standard
colony conditions. Following four days of isolation, midbrain
tyrosine hydroxylase activity was significantly decreased (Figure
14). In addition to these changes, by 16 days the tyrosine hydroxylase in the neostriatum was also found to be significantly increased.
The effect of isolation on tyrosine hydroxylase activity is consistent with the view that one of the consequences of altered adrenergic
transmission is the compensatory regulation of the level of enzyme
activity. That is, since it has been shown that isolation initially
results in a lowering in brain NE levels (Yuwiler, 1972), the
consequent probable decrease in noradrenergic neurotransmission could
be responsible for triggering an increase in tyrosine hydroxylase
activity. The increased tyrosine hydroxylase activity might then be
responsible for the observed subsequent increase in brain NE (Yuwiler,
1972).

    In order to examine the possible functional significance of
these enzymatic changes, spontaneous behavioral activity and amphe-
tamine-induced excitation were compared in isolated and grouped rats.
Rats were isolated or grouped for 4-6 days at which time their gross
locomotor activities (crossovers) were measured for one-hour intervals
in experimental chambers described previously (Segal and Mandell,
1970). D-amphetamine sulfate (1.0 mg/kg, ip) was injected in half
the isolated and grouped rats immediately prior to testing. The
results show that the isolated rats were significantly more active
than the grouped controls (Table 2). Comparison of rearing scores
revealed a similar relationship. These findings are consonant with
our previous report of the relationship between reserpine induced
changes in brain tyrosine hydroxylase activity and behavioral excitation (Segal et al., 1971). That is, both chronic reserpine treatment
and isolation resulted in an increase in midbrain and striatal
enzyme activity concomitant with an increase in spontaneous locomotor activity. The apparent additive effects of isolation and
amphetamine on behavioral activity (Table 1) explain the observation that isolated mice and rats (Katz and Steinberg, 1970) are
more sensitive to toxic effects of stimulant drugs. The decrease in
nerve ending tryptophan hydroxylase may reflect a decrease in the

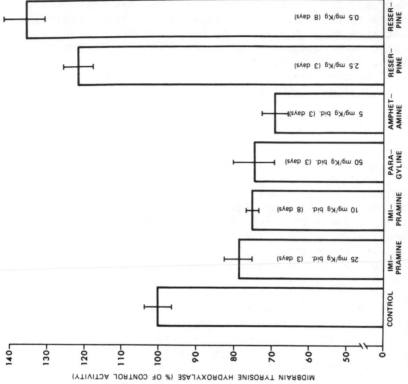

Figure 15. A summary of the effects of chronic drug treatment on the activity of midbrain tyrosine activity. Reserpine given in daily doses of 2.5 mg/kg for 3 days, 0.5 mg/kg for 8 days; imipramine, 25 mg/kg b.i.d. for 3 days, 10 mg/kg b.i.d. for eight days; pargyline, 50 mg/kg b.i.d. for 3 days; amphetamine, 5 mg/kg b.i.d. for 3 days. Note that whereas reserpine produced a significant increase in tyrosine hydroxylase, antidepressant and stimulant drugs lead to a decrease.

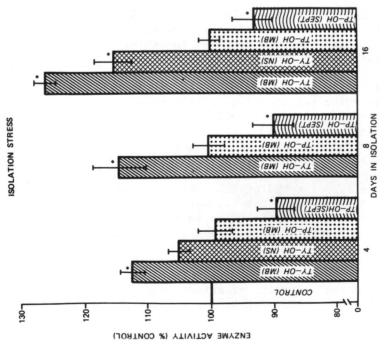

Figure 14. The effect of isolation of rats in a sound attenuated-darkened room for 4, 8, and 16 days on midbrain (MB), neostriatal (NS), and septal (SEPT) tyrosine hydroxylase (TH-OH) and tryptophan hydroxylase (TP-OH) activities. Note that there is an immediate and progressive increase in midbrain tyrosine hydroxylase, and delayed increase in neostriatal tyrosine hydroxylase, and an immediate decrease in septal tryptophan hydroxylase.

TABLE II

Effect of environmental deprivation on spontaneous and amphetamine induced motor activity. Rats were injected with either saline or D-amphetamine sulfate (1.0 mg/kg, ip) immediately prior to the 1 hr. test session.

TREATMENT	ACTIVITY SCORES*
Grouped + saline	54 ± 10
Isolated + saline	$93 \pm 11^{\#}$
Grouped + amphetamine	$227 \pm 20^{+}$
Isolated + amphetamine	$309 \pm 24^{\#+}$

* crossovers (mean ± SEM) for the last 50 min. of 1 hr. session (N = 7 per group)

\# differs from corresponding grouped controls p < .05 (Mann-Whitney-U)

+ differs from corresponding saline controls p < .01 (Mann-Whitney-U)

functional serotonin pool, and therefore a decrease in serotonergic neuronal transmission. Our finding that isolated animals showed an increase in spontaneous activity with a corresponding decrease in tryptophan hydroxylase activity is consonant with previous findings showing an inverse relationship between serotonin levels and gross spontaneous activity (Appel et al., 1970). It is of interest that in addition to isolation, two stimulant drugs, methamphetamine and cocaine also resulted in decreased levels of septal but not of mid-brain tryptophan hydroxylase when acutely administered (Knapp and Mandell, in preparation). Therefore, the increased sensitivity of amphetamine in isolated animals may be due, at least in part, to an additive decrease in nerve ending tryptophan hydroxylase.

d. Morphine. Changes in nerve ending tryptophan hydroxylase activity in response to chronic morphine treatment also appears to be consistent with a compensatory model. As noted previously, Figure 5 shows that septal tryptophan hydroxylase activity in rats treated acutely with 10 mg/kg of morphine was depressed 3 hours after exposure to the drug. However, with chronic subcutaneous injections or with morphine pellet implantation for 3-5 days the nerve ending enzyme levels were significantly elevated. The transient decline appears due to a direct inhibition of tryptophan hydroxylase by morphine which is concentrated by a nerve ending uptake mechanism (Scrafani et al., 1969; Navon and Lajtha, 1970). The delayed enzyme increase would reflect a compensatory response to the consequences of such a decrease (i.e., drop in 5-HT levels and subsequent impairment in serotonergic transmission).

(2) Treatments which correlate with facilitated transmission.
a. Potentiators of central adrenergic synapses. As with the sympathetic nervous system, it can be reasoned that if changes in the amount or activity of the neurotransmitter biosynthetic enzymes in the brain reflect a homeostatic or compensatory mechanism, then those drugs which facilitate transmission ought to produce a consequent decrease in enzyme activity. In order to test this hypothesis, rats were injected, with various chronic regimens, with either (1) imipramine, which prevents the uptake inactivation of catecholamines; (2) pargyline, which appears to prevent inactivation either by MAO inhibition or uptake; or (3) amphetamine which may both release CAs and prevent their inactivation by reuptake.

Figure 15 shows that imipramine at doses of 25 mg/kg (2x day) for three days and at 10 mg/kg (2x day) for eight days produced significant decreases in midbrain tyrosine hydroxylase activity. In addition, pargyline given at a dose of 50 mg/kg (2x day) for three days and 5 mg/kg of amphetamine (2x day) administered for three days also produced a similar decline in the enzyme activity. The reserpine-induced increase in midbrain enzyme activity is presented for the purpose of comparison with these results. As can be seen, there is approximately a two-fold difference in activity between the chronic reserpine and chronic imipramine group.

b. Genetic strains of rats characterized by rate of spontaneous motor activity. Our group has examined the relationship between the levels of regional brain TH and spontaneous behavioral activity in five genetic strains of rats (inbred for a minimum of forty generations) in order to determine if these two variables co-vary. Justification of the choice of spontaneous behavioral activity as an indicator of the functional state of central catecholamine transmission has been outlined elsewhere (Segal and Mandell, 1970). To obtain baseline levels of spontaneous activity, rats were habituated to the experimental chambers and on a subsequent day their level of ambulation was automatically recorded. The results of these behavioral measurements revealed a rather wide interstrain range of activity levels (Figure 16), the BUF strain showing the greatest number of crossovers, while the F344 strain exhibited the least. The striatal and midbrain TH levels were also obtained for the various strains (Figure 16). These results show that the F344 strain, which was the least behaviorally active of the six strains tested, had the highest level of enzyme activity. Conversely, the BUF strain,

FIGURE 16

STRAIN DIFFERENCES IN LEVELS OF TYROSINE HYDROXYLASE
AND SPONTANEOUS ACTIVITY

A. SPONTANEOUS ACTIVITY

Strain		BUF	SD	LEW	ACI	BN	F344
		137*	121	114	92	59	46
		±29	±27	±24	±29	±15	±4
		(n=7)	(n=7)	(n=7)	(n=7)	(n=7)	(n=7)
BUF	137 ± 29	–	–	–	–	–	–
SD	121 ± 27	N.S.	–	–	–	–	–
LEW	114 ± 24	N.S.	N.S.	–	–	–	–
ACI	92 ± 29	N.S.	N.S.	N.S.	–	–	–
BN	59 ± 15	<.01	<.05	<.05	N.S.	–	–
F344	46 ± 4	<.01	<.001	<.001	<.05	N.S.	–

B. STRIATAL TYROSINE HYDROXYLASE

Strain		BUF	SD	LEW	BN	ACI	F344
		12485†	13919	14867	15079	14472	17944
		±556	± 406	± 265	± 982	± 902	± 418
		(n=10)	(n=15)	(n=15)	(n=15)	(n=10)	(n=15)
BUF	12485 ± 556	–	–	–	–	–	–
SD	13919 ± 406	<.01	–	–	–	–	–
LEW	14867 ± 265	<.001	<.01	–	–	–	–
BN	15079 ± 982	N.S.	N.S.	N.S.	–	–	–
ACI	15572 ± 902	<.01	<.05	N.S.	N.S.	–	–
F344	17944 ± 418	<.001	<.001	<.001	<.05	<.002	–

FIGURE 16 (continued)

C. MIDBRAIN TYROSINE HYDROXYLASE

Strain		BUF	ACI	SD	LEW	BN	F344
		1596†	1707	1830	1831	2432	2613
		±139	± 75	±122	±120	±100	± 66
		(n=5)	(n=10)	(n=15)	(n=10)	(n=5)	(n=5)
BUF	1596 ± 139	-	-	-	-	-	-
ACI	1707 ± 75	N.S.	-	-	-	-	-
SD	1830 ± 122	N.S.	N.S.	-	-	-	-
LEW	1831 ± 120	N.S.	N.S.	N.S.	-	-	-
BN	2432 ± 100	<.004	<.001	<.01	<.01	-	-
F344	2613 ± 66	<.004	<.001	<.01	<.001	N.S.	-

Levels of significance were calculated using the Mann-Whitney U test (Siegel, 1957).

*Mean number of cross-overs ± S.E.M.

†Mean net cpm ^3H_2O released/20 min/mg protein ± S.E.M.

which exhibited the highest number of crossovers, had the lowest level of enzyme. The strains intermediate in activity showed a similar but less marked inverse relationship. The negative correlation between behavior and striatal and midbrain tyrosine hydroxylase for the six strains was found to be highly significant. If, in fact, behavioral activity is directly related to the state of adrenergic transmission, then these strain differences are consistent with the previous studies which showed that the activity of tyrosine hydroxylase appeared to be inversely related to the activity of central catecholamine synapses. This work is discussed in greater detail elsewhere (Segal, Kuczenski and Mandell, 1972). It is of interest that these enzymatic differences may not be secondary to weeks or months of strain in behavior pattern in that they appear to be present in some regions shortly after birth.

(3) <u>Treatments which block the receptor</u>. Another area of
support for a compensatory model stems from the observed relationship
between receptor blockade and CA turnover. That is, neuroleptics
such as haloperidol and chlorpromazine as well as other agents
believed to block CA receptors have been found to accelerate an α-
methyl-tyrosine-induced CA depletion in the brain (Anden et al.,
1964). This has been interpreted as resulting from the feedback
communication from the blocked postsynaptic receptor to the presynaptic neurons. This results in an increase in presynaptic neuron
activity and a consequent increase in the rate of CA disappearance.
If the nigro-neostriatal dopamine pathway is transected unilaterally
in such an experiment the induced disappearance of DA is observed
only on the intact side. Consonant with the concept of an adaptive
neural feedback triggered by alterations in transmission, is the
recent report that amphetamine produced a decrease in the spontaneous
activity of single cells in the locus ceruleus, a nucleus enriched
in noradrenergic cell bodies (Grahame and Aghajanian, 1971). Most
of the cells examined outside of this area exhibited an increased
discharge rate or no change at all. This suggested that a relatively specific feedback inhibition of locus ceruleus neurons was
initiated by the amphetamine induced facilitation of noradrenergic
transmission of catecholamine nerve endings. Chlorpromazine, which
is believed to have the opposite effect on catecholamine transmission
reversed the amphetamine effect in most cases. It appears that feedback may be an important factor in modulating the presynaptic nerve
activity and its influence on neurotransmitter dynamics. The specific relationship between such feedback and the regulation of the
level and/or activity of neurotransmitter biosynthetic enzymes constitute a major focus for the future of our work. In regard to
these specific "receptor blockers" it is interesting that chronically
administered chlorpromazine (10 mg/kg ip for 8 days) and haloperidol
(0.5 mg/kg, ip for 8 days) produced a significant increase in striatal tyrosine hydroxylase activity.

To summarize the conclusions from studies of chronic pharmacological and nonpharmacological treatments: there is a growing body
of evidence supporting the view that as with the adaptive enzymatic
changes reported to occur in the adrenal medulla and sympathetic
ganglia, brain tyrosine hydroxylase and under some circumstances,
tryptophan hydroxylase, may be responsive to relatively prolonged
alterations in biogenic amine mediated transmission. In addition,
it may be speculated that such alterations in synaptic transmission
are communicated from the postsynaptic to the presynaptic neuron
and that enzymatic changes are modulated by this feedback neural
input. The relevant input for such enzymatic adaptation may be in
the form of action potential or some tonic state changes similar to
the prolonged hyperpolarization produced by SIF cells in the superior cervical ganglia (McAfee et al., 1971).

c. <u>Adaptive changes in central noradrenergic receptors</u>. A growing body of evidence suggests that receptor adjustments may occur in response to long-term alterations in synaptic transmission in peripheral neural systems. Most of this research has involved the examination of effector tissue following various intervals of denervation. Such studies have frequently reported that receptor sensitivity to the physiological transmitter substance progressively increases following the cessation of neural input. Several mechanisms have been proposed as mediating this receptor change including the proliferation of receptor sites and the activation of unmasking of pre-existing sites. At present, there is little evidence for the existence of similar phenomena in the CNS (Potter and Mollinoff, 1972; Changeux et al., 1970).

Recently, we have explored the possibility that alterations in receptor sensitivity may occur in the CNS. In the first such study, we examined the hypothesis that thyroid state may directly influence the sensitivity of adrenergic receptors. Rats, previously implanted with a cannula in the lateral ventricle, were injected with thyroxine (0.4 mg/kg, ip, for 10 days) after which they were infused with NE intraventricularly and their gross motor activity was compared with appropriate controls. It was reasoned that if noradrenergic receptors were sensitized by thyroxine, the behaviorally activating effects of centrally administered NE would be potentiated by treatment with thyroxine. On Day 10, rats were placed in an experimental chamber and allowed to habituate for 1 hour after which they were infused with NE for 1.5 hours. The results indicated that the tyrosine-treated rats were significantly more responsive to NE infusion than were controls (Figure 17). Some evidence also indicated that response of thyroidectomized animals to NE infusion was markedly delayed and diminished as noted in a previous section.

In a study which more closely resembled the usual approach to receptor manipulation in the peripheral nervous system, we selectively destroyed CA terminals with central injections of 6-hydroxydopamine (6-HD). Approximately twenty days following the intraventricular administration of 250 µg of 6-HD, midbrain and striatal TH levels were markedly reduced. The rate of decline in enzyme activity was considerably faster in the striatum than in the midbrain (Figure 18). The disparity in the rate of decline may reflect the difference in proximity to the site of injection or more likely, the retrograde degeneration of catecholamine neurons. The persistent decrease in tyrosine hydroxylase which is paralleled by a relatively specific drop in CA levels (Uretsky and Iverson, 1969) is consistent with a destruction of CA nerve terminals. However, in spite of the fact that brain CAs appear to be markedly and permanently lowered by this treatment, we and others (Burkard et al., 1969) have found that behavioral activity returns to approximately normal levels by 15

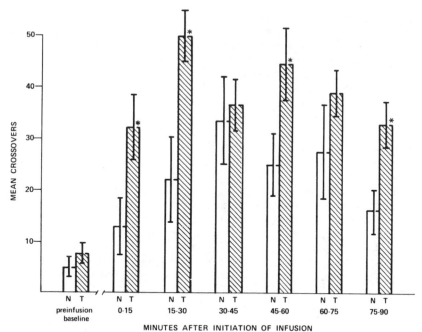

Figure 17. The effects of the infusion of 1.0 µg/ml NE (DL Arterenol-HCl) on the behavioral activity of nontreated (n) and thyroid treated (T) rats. Crossovers were measured in 15 minute blocks for a 2.5 hr period. During the first hour, the animals received no infusion, and the last 15 minute block of this hour was used as the preinfusion base line. The data is presented as the mean number of crossovers ± SEM (brackets). Thyroid treated (T) rats exhibited a potentiated response to NE infusion with respect to the onset, duration and magnitude of the hyperactivity. * = significantly different from controls; p < 0.01.

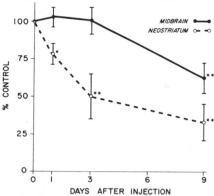

Figure 18. Tyrosine hydroxylase activity at various times after the intraventricular administration of 250 µg 6-hydroxydopamine in 25 µl. *p < 0.01; **p < 0.001 as compared to saline injected controls.

days after the injection of 6-HD. We speculated that this recovery was due to an increased receptor sensitivity to the NE discharged by the neurons which survived the destructive treatment. To test this hypothesis, animals which had been injected intraventricularly with either saline or 250 µg of 6-HD, 20-22 days before, were infused intraventricularly with NE. Pre-infusion tests indicated that there were no significant differences between the saline and 6-HD treated animals; however, the responsivity of the 6-HD group to NE was far greater than that of the controls. This was certainly consistent with the development of supersensitivity by the "denervated" central noradrenergic receptors.

Since adenyl cyclase has been shown to be a norepinephrine-sensive, receptor-like enzyme in peripheral systems (Murad et al., 1962), the activity of this enzyme was examined in a pilot study of the midbrian and striatal regions of rats injected with either saline or 6-HD 20 days prior to sacrifice. The results showed a pronounced elevation in enzyme activity in both brain parts of the 6-HD treated animals. An increase in this enzyme might account for the functional sensitivity to intraventricular NE via a cyclic-AMP mediated post-receptor mechanism.

These results suggest that in addition to the possibility for pre-synaptic enzymatic adaptational mechanisms, receptor adjustments (perhaps in the form of adenylate cyclase alterations) may also play an important role in regulating synapses.

III. SOME THOUGHTS CONCERNING THE INTEGRATIVE ORGANIZATION OF THESE ADAPTIVE CHANGES

Our results thus far have suggested that both presynaptic and post-synaptic macromolecular mechanisms may be involved in the primary mode of action of some experimental treatments as well as operating to maintain the functional components of biogenic amine neural transmission within relatively restricted limits. One group of questions which appear worthy of consideration concern the manner in which these two potential mechanisms may be integrated. It is conceivable that both pre- and postsynaptic adjustments may occur simultaneously. An example of this may be our finding that following chronic reserpine treatment there appears to be a presynaptic compensatory response in the form of an increase in the catechola-mine biosynthetic enzyme, tyrosine hydroxylase. Palmer (1972) on the other hand, has reported a reserpine induced increase in adenyl cyclase activity and has suggested that such changes may represent the molecular mechanism responsible for supersensitivity following impaired neural transmission. Perhaps the supersensitivity to amphetamine after reserpine represents a functional expression of such a change. Although the evidence from such an example suggests the similarity of these two sets of adaptive changes, very careful

multiple measure experiments will be required to rule out differences
in sensitivity, order, and/or magnitude between them.

It is possible that some circumstances would call forth a
preclusive adjustment: An extreme example would be that following
the intracerebral administration of 6-hydroxydopamine, which is
believed to selectively destroy adrenergic terminals, the only
available compensatory mechanism is in the postsynaptic mechanism.
Under these circumstances it might be expected that the magnitude
of the postsynaptic adjustment (as might be indicated by the magnitude of the changes in adenyl cyclase) may be greater than, for
example, following reserpine in which both pre- and postsynaptic
mechanisms would get a chance to contribute to the adaptation.

The effects obtained following thyroidectomy may represent a
situation which would be the reverse of that induced by 6-hydroxy-
dopamine. The receptor sensitivity may be relatively fixed by the
chronic deficiency of thyroid hormone leaving only the presynaptic
enzyme adjustments to compensate for the alterations in transmission.
Obviously any statement regarding the relative contribution of these
two sets of potential regulatory mechanisms must await further
research.

Another set of potentially generative questions involves the
nature of the critical signal which triggers the pre- and postsynaptic adaptational mechanisms. One same possibility is that
intraneuronal changes may be responsible for the alterations in the
rate-limiting biosynthetic enzymes. For example, long-term changes
in the intraneuronal concentration of neurotransmitter may affect
either the physical state of the enzyme and/or its rate of synthesis
and degradation. Our studies to date appear inconsistent with this
position in that there is no consistent relationship between the
direction of change induced in the concentration of brain CAs and
the direction of enzyme change. For example both chronic reserpine
treatment and thyroidectomy result in an elevated regional brain
tyrosine hydroxylase activity, yet reserpine depletes intraneuronal
CA stores while thyroidectomy appears to produce increase in such
stores. The effect common to both of these treatments is their
functional impairment of central adrenergic transmission. As a
consequence of such observations, we suggest that the relevant information leading to adaptive changes in neurotransmitter enzymes
is the functional state of synaptic transmission independent of
the intraneuronal levels of transmitter substance. It should be
acknowledged that intraneuronal events other than changes in the
level of transmitter substances may be critical. For example,
changes in ion or precursor levels may be found to be more closely
related to the chronic treatment induced enzyme alterations. In
fact, even the level of transmitter substance cannot be ruled out
on the basis of available evidence since there may be specific pool

brain concentrations. In spite of these reservations, our current orientation could be called "the law of synaptic effect" in which the functional result of synaptic transmission is the prime determinant of pre- and postsynaptic adaptations and not the concentration or state of any one of the elements of adaptation.

At the present time it appears that a very likely mechanism for modulating biosynthetic potential involves interneuronal feedback to the presynaptic neuron, the relevant stimulus being the state of transmission. Several alternatives exist as to the characteristics of the electrical input controlling the rate of enzyme production. One possibility is that the synthesis of the enzyme is directly related to the rate of neuronal firing. The feedback input would simply contribute to and algebraically summate with other inputs impinging upon the presynaptic neuron. Another rather intriguing possibility is that the feedback neural input has as its specific unique function that of controlling the rate of enzyme synthesis; a "trophic" regulating system that operated apart from regulators of firing rate. This could be accomplished either by the activation of a specialized site on the postsynaptic membrane or by the specific nature of the controlling input. An example of the latter type of input might be the slow hyperpolarization associated with the dopamine mediated SIF cells of the superior cervical ganglion (McAfee et al., 1971). The consequences produced by some parameter of this phasic change in excitabilty state (e.g., duration, magnitude and/or frequency of occurrence) may be causally related to some of the macromolecular adaptive mechanisms described in this paper. We are currently in the process of exploring some of these possibilities.

Another issue, which is perhaps related, concerns the regulating information to the receptor. Receptor sensitivity may be modulated by several mechanisms. One possibility is that a specific trophic substance is released along with the transmitter substance. In this case, the regulator of the sensitivity of the receptor may be different than the depolarizing transmitter substance. Receptor sensitivity may also be modulated by the interaction of the transmitter substance at either the site at which it produces its agonistic action or at a second specialized site on the membrane which functions to set the gain for the agonistic site. Consonant with this latter model is the accumulating evidence which indicates that there may be two functionally different binding sites associated with receptor membranes. We are currently in the process of applying immobilization techniques to the 6-hydroxydopamine denervated animal brains in search of alterations in adrenergic "receptor" molecules.

IV. SOME IMPLICATIONS FOR A PATHOPHYSIOLOGY OF MENTAL DISEASE

One clinical implication of the synaptic adaptive mechanisms elucidated by the work discussed above involves their possible roles in an understanding of a pathophysiology of mental disorder. There have been two major themes in biological psychiatry in recent years to which considerable complexity must be added in light of the adaptive mechanisms described in this paper: the first involves the theory that the schizophrenias are caused by endogenous psychotoxins such as hallucinogens; the second relates extremes of mood and affect disorder to a state in the brain of "too much" or "too little" mood transmitter. Thus, research efforts were directed to the search for a chemical condition whose presence would be both necessary and sufficient for the expression of mental disease in man (either an abnormal compound or too much or too little of a normally present compound). Like many of the primitive phases of research in other medical disorders, the search for the hypothetical "schizococcus" in one form or another has dominated biological research in psychiatry for a number of years. As our understanding of the pathophysiology of various medical and neurological diseases changed, the character of this schizococcus changed. Instead of being the result of the influence of a foreign agent, it was seen as an indigenous agent that was part of the body's normal flora and excreted psychotoxins or influenced "levels" on a continuing basis. With the discovery of the hereditary metabolic diseases and their associated missing or defective enzymes, the psychotoxic agent became the aberrant product of a previously minor biochemical pathway. After this and following the popularity of the autoimmune concept of disease, the agent became an antibody that altered the function of brain structures or metabolic pathways. The new era of hallucinogen synthesis and discovery led to the speculation that one or another of these potent compounds could be made from at least superficially naturally occurring compounds in man under some special hereditary or environmental circumstances.

Our laboratory has recently elucidated an enzyme in human brain that can convert serotonin, a neurotransmitter, to a centrally active hallucinogen-like compound, bufotenine or tryptamine to DMT (Morgan and Mandell, 1969; Mandell and Morgan, 1971; Mandell et al., 1971). It is called indole(ethyl)amine N-methyltransferase (IENMT) and has been purified some 30-fold from sheep brain homogenate using tryptamine as the indoleamine methyl acceptor. A similar enzyme has recently been reported from rat brain by Saavedra and Axelrod (1972) that can make dimethyltryptamine (DMT) from the tryptamine. The possibility for a pathophysiology involving this enzyme may be suggested in which a defective or chemically inhibited monoamine oxidase could lead to an indoleamine methylation shunt and the synthesis of abnormal centrally active compounds. A recent series of studies of multiple forms (Youdin, 1972) of human brain monoamine oxidase and their regional and substrate specification suggest that a relative indoleamine monoamine oxidase deficiency may be present compared to changes in the three other forms.

Although probably more rational than some other pathophysiologies in psychiatry, the search for a single etiological agent does not seem consonant with the findings reviewed in this paper. It has been clear that even in such relatively straightforward medical diseases as tuberculosis, such factors as hereditary predisposition, race, nutritional state, and emotional factors play some role in the final expression of the disease. Phenylketonuria, perhaps the most classical of the well-characterized hereditary metabolic diseases that affect brain function, is still surrounded by spirited debate about its pathophysiology (Man and Spooner, 1968). Considerable disagreement exists as to why the behavioral and functional syndromes may vary so much among patients with the same metabolic defect. In view of these and other considerations, the characterization of the body's mechanisms of defense and determinants of vulnerability to pathological agents have increasingly become the focus of study of scientists studying the pathophysiology of disease. The immune system and its potential for inactivating abnormal substances, for example, has become a major focus for cancer research. Examination of the research frontiers in other diseases reflect the shift in attention from the character of the etiological agent to the status of the host's defenses. This sort of shift in focus had already occurred in psychodynamic thinking in psychiatry decades ago. Traumatic life events as etiological agents had given way to a focus of the therapist-theorist on the "coping-style" of the individual. In the psychobiological arena until very recently, we have been without a vocabulary of adaptive phenomena in the brain which was sufficiently well defined to be invoked in discussion of the role of the brain's "defenses" in biological terms. Our studies have demonstrated a complex and interacting system of adaptive devices that appears to indicate that a psychotoxic agent alone could not be the cause of the chronic mental disorders. A hallucinogen-like product of a brain enzyme could not in and itself produce a long-lasting excitatory or inhibitory disruptive influence in brain function. The wide variety of compensatory changes that such agents most probably would trigger under normal circumstances would prevent this from happening.

Similar considerations also raise concerns about the acceptance of a pathophysiological model involving a deficiency or excess of neurotransmitter in the affect disorders. This kind of approach might be represented by the "catecholamine theory of mood." A failure in synaptic adaptive mechanisms appears to be necessary in order to account for a prolonged abnormality in neurotransmitter function. The behavioral recovery in animals whose norepinephrine nerve endings were 60–80% destroyed by the intraventricular administration of 6-hydroxydopamine required about two weeks and was associated with a marked functional supersensitivity of the catecholamine receptors. It would seem that the deficiency of functional catecholamines hypothesized to be a necessary neurochemical

correlate of the clinical depressive illnesses should certainly be
compensated for by similar receptor alterations and/or by the reflex
changes in catecholamine biosynthetic enzyme activity.

The mechanisms that have been elucidated above appear to have
as their major adaptive function the maintenance of a specific range
of neurotransmitter function. As with tuberculosis, one must posit
a failure in the host's defenses as well as the presence of the
agent in order to approach an understanding of the pathophysiology
of mental disease. Whether some etiological agents have multiple
capacities so that in addition to their primary disruptive effects
they also alter the capacity of central synaptic compensatory
mechanisms has not been demonstrated for any neurometabolic disease
except those that actually alter CNS tissue composition. Even this
type of alteration (for example, the return to relatively normal
behavior by the rat following the destruction of catecholamine
nerve endings by 6-hydroxydopamine) may be adequately compensated.

This line of thinking has suggested that individual differences
in susceptibility to mental disorder may be due to genetically
linked responsivity to synaptic adaptive mechanisms. This led us
to search for appropriate animals specifically for genetic differences in the metabolic adaptive capacity of biogenic amine central
synapses in some strains. As noted above, we have recently demonstrated that five strains of rats that differed markedly in spontaneous exploratory behavior, differed also in a systematic way in the
regional specific activity of the neurotransmitter biosynthetic
enzyme involving the norepinephrine and dopamine pathways. This has
also been true for recent work with serotonin and acetylcholine
biosynthetic enzymes. We are beginning to investigate the problem
of the differential capacity and rate of development of tolerance
to a variety of centrally active agents in these various strains.
Could it be that a chemical model of mental disease should feature
genetic differences in metabolic adaptive capacity to both excitation
and inhibition in central synaptic mechanisms? Perhaps people that
get mentally ill are those that when faced with aberrant amounts or
kinds of chemical messengers (in kind or amount) cannot adapt sufficiently. Griffith et al. (1972) have shown that whereas all his
experimental subjects eventually developed classical paranoid
schizophrenia when he gave them 10 mg of amphetamine every hour, the
range of the individual differences in resistance to the syndrome
was marked. One subject became paranoid in hours, others took days.
Could these findings be a reflection of both aspects of a two-
factor chemical theory of psychosis involving both a toxic agent
and an adaptational failure? As noted previously, one could
hypothesize that following the inhibition of MAO by high levels of
the amphetamines there was the production of hallucinogenic N,N-
dimethylated indoleamines such as bufotenine or dimethyltryptamine.
In addition, one could speculate that the individual differences in

the development of psychosis to the same psychotoxin was a function
of the metabolic adaptive potential of the brain's pre- and postsynaptic mechanisms for normalizing excitability.

It is interesting that Freud began his early neurobiological
theorizing with a toxin-like concept of "instinct" (especially
dammed-up instinct) which he felt played a major role in disrupting
personality function and causing mental disease. Later he began to
consider the importance of the inhibitory functions of the ego in
its capacity to dampen uncontrolled excitation and delay its discharge. The Ego Psychology of the post-Freudians and Neo-Freudians
has focused most of its metapsychological theory on adaptive
capacity and control of excitation and impulse as predictors of
success in life and resistance to mental illness. Even the recent
experimental and research psychophathologists have emphasized this
dimension. Broen and Storms (1966) have derived a pathophysiological
theory of schizophrenia that posits a defect characterized by a
failure in ability to dampen excitability states producing over-
generalization in perception and cognition and thus leading to the
characteristic thought disorder. Following a group of Danish
children for years who were genetically loaded for schizophrenia
(mothers were schizophrenic), Dr. Sarnoff Mednick has been able to
discriminate those that will get sick from those that won't on the
basis of their individual capacity to extinguish autonomic responses
to repeated startling stimuli. Those that become ill, revealed
their potential years before by their inability to habituate their
GSR. Recently, Ban, Lehman and Green (1970) evolved a diagnostic
battery for a range of patients with mental disorders which includes
a subgroup of schizophrenics whose response to novel stimuli (Orient-
ing Response) failed to habituate with repeated trials. It is
tempting to speculate that this failure to develop stability in
central synaptic systems may involve a defect in some of the adaptational processes that we have been examining.

The two most popular current pathophysiological themes in
Biological Psychiatry may be open to serious question, at least in
their most simple forms, if the central synaptic adaptive mechanisms
prove to be as effective as they appear to be in work reported in
this paper. The psychotogen theory of schizophrenia (like the
quickly developing tachyphylaxis to LSD with cross tolerance to
other hallucinogens) is not a defensible experimental set when both
pre- and postsynaptic mechanisms appear prepared to act in concert
to dampen the effect of the disruption. The "too much" or "too
little" neurotransmitter theory of affect disorder is equally difficult to accept in the presence of normally functioning synaptic
adaptive mechanisms. It is clear that the next model posited in
Biological Psychiatry will require both an explanation of the primary
disruption in normal CNS function as well as the failure of the normally effective adaptive mechanisms. It is possible that impairment

could be seen as either an alteration in the appropriate signal for adaptive change or in the limited capacity of the system to make the appropriate change. Thus far we have little in the way of an experimental example of such theoretical constraints at work outside of the examples of the limited adaptive capacity on either side of the synapse in the 6-hydroxydopamine or thyroidectomy experiments described above. Combinations of primary and secondary pathophysiologies in which adaptational failure as well as primary disruptions must be explained would certainly be the next order of theory in Biological Psychiatry. It promises to be both more difficult but at the same time probably more complete.

A second, more fanciful implication of this work involves the possible biological basis of a treatment strategy aimed at habituating to or "working through" sources of pathological excitation in patients rather than shielding or protecting them. In light of our work, it is tempting to suggest that protective strategies of treatment involving reductions in interpersonal activity may increase the potential trauma of later interpersonal stresses. Chronic strategies of treatment involving increasing doses of significant central excitation may serve to facilitate appropriate synaptic adaptive mechanisms and aid the development of resistance to disruptive stimuli. The sensory deprived animals in our studies developed an increase in the rate-limiting catecholamine bio-synthetic enzyme and increased sensitivity to the amphetamines. In addition to "working through" specific stimuli, it is possible that more general manipulations of the synaptic adaptive mechanisms could take place via exercise, heat, cold and other similar physiological experiences.

It is tempting to conclude that we are in the Second Generation of explanation and theory in Biological Psychiatry. The first was in terms of causative agents (e.g., hallucinogen, deficiency of catecholamines). The second involves the beginning of an understanding of the brain's adaptive processes.

ACKNOWLEDGEMENTS

This work was supported by NIMH Grants #MH 14360 and MH 18065 as well as the Friends of Psychiatric Research of San Diego, Inc.

REFERENCES

Anden, N.E., Roos, B.-E., and Werdinius, B. 1964. Effects of chlorpromazine, haloperidol and reserpine on the levels of phenolic acids in rabbit corpus striatum. Life Sci., 3, 149.
Appel, J.B., Lovell, R.A. and Freedman, D.X. 1970. Alterations in the behavioral effects of LSD by pretreatment with p-Chlorphenylalanine and α-Methyl-p-Tyrosine. Psychopharmacologia, 18, 387.

Axelrod, J., Mueller, R.A., Henry, J.P. and Stephens, P.M. 1970.
  Changes in enzymes involved in the biosynthesis and metabolism
  of noradrenaline and adrenaline after psychosocial stimulation.
  Nature, 225, 1059.
Ben, T.A., Lenman, H.E. and Green, A.A. 1970. Conditioned reflex
  variables in the prediction of therapeutic responsiveness to
  phenothiazines in the schizophrenias. In: Psychopharmacology
  and the Individual Patient, eds. J.K. Wittenborn, S.C. Goldberg
  and P.R.A. May, Raven Press, New York, pp. 55-71.
Barondes, S.H. 1970. Cerebral protein synthesis inhibitors block
  long-term memory. Internat. Rev. Neurobiol., 12, 177-203.
Boyaner, H.G. and Radouco-Thomas, S. 1971. Partial antagonism by
  exogenous calcium of the depressant effect of reserpine in rat
  shuttle box behavior. Brain Res., 33, 589-591.
Broen, W.E. and Storms, L.H. 1966. Lawful Disorganization: The
  process underlying a schizophrenic syndrome. Psychol. Rev. 73,
  265.
Burkard, W.P., Jaffe, M. and Blum, J. 1969. Effect of 6-hydroxy-
  dopamine on behavior and cerebral amine content in rats.
  Experientia, 25, 1295.
Changeux, J.P., Kasai, M. and Lee, C.Y. 1970. Use of a snake venom
  toxin to characterize the cholinergic receptor protein. 1970.
  Proc. Nat Acad. Sci., 67, 1241-1249.
Cheney, D.L., Goldstein, A., Algui, S. and Costa, E. 1971. Narcotic
  tolerance and dependence: Lack of relationship with serotonin
  turnover in brain. Science, 171, 1169.
Costa, E. and Neff, N.H. 1970. Estimation of turnover rates to
  study the metabolic regulation of the steady-state level of
  neuronal monoamines. Handbook of Neurochemistry. IV., ed. Abel
  Lajtha, Plenum Press, New York, pp. 45-90.
Dahlstrom, A. and Fuxe, K. 1964. Evidence for the existence of
  monoamine-containing neurons in the central nervous system. Acta
  Physiol. Scand. 62 (Suppl. 232), 1.
Dairman, W. and Udenfriend, S. 1970. Increased conversion of
  tyrosine to catecholamines in the intact rat following elevation
  of tissue tyrosine hydroxylase levels by administered phenoxyben-
  zamine. Mol. Pharm. 6, 350-356.
Emlen, W., Sega., D.S. and Mandell, A.J. 1972. Thyroid state:
  Effects of pre- and postsynaptic central noradrenergic mechanisms.
  Science, 175, 49.
Fahn, S., Rodman, J.S. and Cole, L.J. 1969. Association of tyrosine
  hydroxylase with synaptic vesicles in bovine nucleus. J.
  Neurochem.,16, 1293.
Fernstrom, J.D. and Wurtman, R.J. 1971. Brain serotonin content:
  Physiological dependence on plasma tryptophan levels. Science,
  173, 149.
Gal, E. M., Roggeveen, A.E. and Millard, S.A. 1970. D,L-[2^{14}C]
  p-chlorophenylalanine as an inhibitor of tryptophan 5-hydroxylase.
  J. Neurochem., 17, 1221.

Grahame, A.W. and Aghajanian, G.K. 1971. Effects of amphetamine on single cell activity in a catecholamine nucleus, the locus coerleus. Nature, 234, 100-102.

Grahame-Smith, D.G. 1967. The biosynthesis of 5-hydroxytryptamine in brain. Biochem. J., 105, 351.

Grahame-Smith, D.L. and Parfitt, A.N. 1970. Tryptophan transport across the synaptosomal membrane. J. Neurochem., 17, 1339.

Grahame-Smith, D.J. 1971. Studies in vivo on the relationship between brain tryptophan, brain 5-HT synthesis and hyperactivity in rats treated with a monoamine oxidase inhibitor and 1-tryptophan. J. Neurochem., 18, 1053.

Griffith, J.D., Cavanaugh, J., Held, J. and Oates, J.A. 1972. Dextroamphetamine: Evaluation of psychomimetic properties in man. Arch. Gen. Psychiat., 26, 97.

Ichiyama, A., Nakamura, S., Nishizuka, Y. and Hayaishi, O. 1970. Enzymic studies on the biosynthesis of serotonin in mammalian brain. J. Biol. Chem., 245, 1699.

Jequier, E., Lovenberg, W. and Sjoerdsma, A. 1967. Tryptophan hydroxylase inhibition: The mechanism by which p-chlorophenylalanine depletes rat brain serotonin. Mol. Pharmacol., 3, 244.

Katz, D.M. and Steinberg, H. 1970. Long-term isolation in rats reduces morphine response. Nature, 228, 469.

Knapp, S. and Mandell, A.J. 1972. Parachlorophenylalanine - its three phase sequence of interactions with the two forms of tryptophan hydroxylase. Submitted for publication.

Knapp, S. and Mandell, A.J. 1972b. Some drug effects of the functions of the two physical forms of tyrptophan-5-hydroxylase: Influence on hydroxylation and uptake of substrate. Conference on Serotonin and Behavior, Palo Alto, ed. Barchas, J. January, 1972, in press.

Knapp, S. and Mandell, A.J. In preparation, 1972 a and b.

Kopin, I.J., Breese, G.R., Krauss, K.R. and Weise, V.K. 1968. Selective release of newly synthesized norepinephrine from the cat spleen during sympathetic nerve stimulation. J. Pharm. Exp. Therap., 161, 271-278.

Krishna, G., Hynie, S. and Brodie, B.B. 1968. Effects of thyroid hormones on adenyl cyclase in adipose tissue and on free fatty acid mobilization. Proc. Nat. Acad. Sci. 89, 884.

Kuczenski, R.T. and Mandell, A.J. 1972a. Regulatory properties of soluble and particulate rat brain tyrosine hydroxylase. J. Biol. Chem., in press (May, 1972).

Kuczenski, R.T. and Mandell, A.J. 1972b. Allosteric activation of hypothalamic tyrosine hydroxylase by ions and sulfated mucopolysaccharides. J. Neurochem., 19, 131.

Kuczenski, R.T. and Mandell, A.J. Brain tyrosine hydroxylase: Activation by trypsin incubation. Submitted for publication, 1972.

Kvetnansky, R., Weise, V. and Kopin, I. Elevation of adrenal tyrosine hydroxylase and phenylethanolamine-N-methyl transferase by repeated immobiliation of rats. Endocrinology, 87, 744.

Louis, W.J., Spector, S., Tabei, R. and Sjoerdsma, A. 1969. Synthesis and turnover of norepinephrine in th heart of the spontaneously hypertensive rat. Circulat. Res. 24, 85.

Mandell, A.J., Buckingham, B. and Segal, D. 1971. Behavioral, metabolic and enzymatic studies of a brain indole(ethyl) amine N-methylating system. In: Brain Chemistry and Mental Disease eds. B.T. Ho and W. McIssac, Plenum Press, New York, pp. 37-60.

Mandell, A.J., Knapp, S. and Kuczenski, R. 1971. An amphetamine induced shift in the subcellular distribution of caudate tyrosine hydroxylase. Proc. 33rd Meeting Drug Abuse Committee, Nat'l. Res. Council--Nat'l Acad. Sci., 1971, pp. 742-766.

Mandell, A. J., Knapp, S., Kuczenski, R.T. and Segal, D.S. 1972. A methamphetamine induced shift in the physical state of rat caudate tyrosine hydroxylase. Biochem. Pharmacol., in press.

Mandell, A.J. and Morgan, M. 1971. An Indole(ethyl)amine N-methyltransferase in human brain. Nature, 230, 85.

Mandell, A.J. and Spooner, C.E. 1968. Psychochemical research studies in man. Science, 162, 1442.

McAfee, D.A., Schorderet, M. and Greengard, P. 1971. Adenosine 3',5'-monophosphate in nervous tissue: Increase associated with synaptic transmission. Science, 171, 1156.

McGeer, P.L., Bagchi, s.P. and McGeer, E.G. 1965. Subcellular localization of tyrosine hydroxylase in beef caudate nucleus. Life Sciences, 4, 1859-1867.

Molinoff, P.B., Brimijoin, S., Weinshilboum, R. and Axelrod, J. 1970. Neurally mediated increase in dopamine β-hydroxylase activity. Proc. Nat. Acad. Sci. 66, 453.

Morgan, M. and Mandell, A.J. 1969. Indole(ethyl)amine N-methyltransferase in the brain. Science, 165, 492.

Mueller, R.A., Thoenen, H. and Axelrod, J. 1969a. Increase in tyrosine hydroxylase activity after reserpine administration. J. Pharmacol. Exp. Therap., 169, 74-79.

Mueller, R.A., Thoenen, H. and Axelrod, J. 1969b. Adrenal tyrosine hydroxylase: Compensatory increase in activity after chemical sympathectomy. Science, 163, 468.

Murad, F., Chi, Y.-M., Rall, T.W. and Sutherland, E.W. 1962. Adenyl cyclase III. The effect of catecholamines and choline esters on the formation of adenosine 3',5'-phosphate by preparations from cardiac muscle and liver. J. Biol. Chem., 237, 1233.

Navon, S. and Lajtha, A. 1970. The uptake of morphine into synaptosomes. Brain Res., 24, 534.

Palmer, G.C. 1972. Increased cyclic AMP response to norepinephrine in the rat brain following 6-hydroxydopamine. Neuropharmacology, 11, 145.

Peters, D.A., McGeer, P.L. and McGeer, E. 1968. The distribution of tryptophan hydroxylase in cat brain. J. Neurochem., 15, 1431.

Potter, L. and Mollinoff, P.B. 1972. Perspectives in Neuropharmacology, ed., S.H. Snyder, Oxford University Press, London, in press.

Prange, A.J., Meek, J.L. and Lipton, M.A. 1970. Catecholamines:

Diminished rate of synthesis in rat brain and heart after thyroxine pretreatment. Life Sci. 1, 901.

Prange, A.J., Wilson, I.C., Rabon, M. and Lipton, M.A. 1969. Enhancement of imipramine antidepressant activity by thyroid hormone. Amer. J. Psychiat. 126, 457.

Saavedra, J.M. and Axelrod, J. 1972. Psychomimetic N-methylated tryptamines: Formation in brain in vivo and in vitro. Science, 175, 1365.

Schildkraut, J.J. 1970. Neuropsychopharmacology and the Affective Disorders. Little, Brown & Company, Boston.

Schildkraut, J.J. and Kety, S.S. 1967. Biogenic amines and emotion. Science, 156, 21.

Scrafani, J.T., Williams, N. and Clonet, D.H. 1969. Binding of dihydromorphine to subcellular fractions of rat brain. The Pharmacologist, 11, 256.

Segal, D.S., Kuczenski, R.T. and Mandell, A.J. 1972. Strain differences in behavior and brain tyrosine hydroxylase activity. Behavioral Biol., 7, 75.

Segal, D.S. and Mandell, A.J. 1970. Behavioral activation of rats during intraventricular infusion of norepinephrine. Proc. Nat. Acad. Sci. 66, 289.

Segal, D.S., Sullivan, J.L., Kuczenski, R.T. and Mandell, A.J. 1971. Effects of long-term reserpine treatment on brain tyrosine hydroxylase and behavioral activity. Science 173 847-849.

Siegel, S. 1956. Nonparametric Statistics for the Behavioral Sciences. McGraw-Hill Book Co., New York.

Sullivan, J.L., Segal, D.S., Kuczenski, R.T. and Mandell, A.J. 1972. Propranolol induced rapid activation of rat striatal tyrosine hydroxylase concomitant with behavioral depression. Biol. Psychiat. 4, 193.

Tarver, J., Berkowitz, B. and Spector, S. 1971. L-dopa-alterations in tyrosine hydroxylase and nonoamine oxidase activity in blood vessels. Nature New Biology, 231, 252.

Thoenen, H. 1970. Induction of tyrosine hydroxylase in peripheral and central adrenergic neurones by cold-exposure of rats. Nature, 228, 861.

Tozer, T.N., Neff, N.H. and Brodie, B.B. 1966. Application of steady state kinetics to the synthesis rate and turnover time of serotonin in the brain of normal and reserpine-treated rats. J. Pharmacol. Exp. Therap., 153, 177.

Uretsky, N.J., Iverson, L.L. 1969. Effects of 6-hydroxydopamine on noradrenaline-containing neurones in the rat brain. Nature 221, 557.

Vos, J., Kuriyama, K. and Roberts, E. 1968. Electrophoretic mobilities of brain subcellular particles and binding of -aminobuteric acid, acetylcholine, norepinephrine, and 5-hydroxytryptamine. Brain Res., 9, 224.

Way, E.L., Foh, H.H. and Shen, F. 1968. Morphine tolerance, physical dependence, and synthesis of brain 5-hydroxytryptamine.

Science 162, 1290.

Weiner, N. 1970. Regulation of norepinephrine biosynthesis. Ann. Rev. Pharmacol. 10, 273.

Whitsett, T.L., Halushka, P.V. and Goldberg, L.I. 1970. Attenuation of postganglionic synpathetic nerve activity by L-dopa. Circulat. Res. 27, 561.

Wurzburger, R.J. and Musacchio, J.M. 1971. Subcellular distribution and aggregation of bovine adrenal tyrosine hydroxylase. J. Pharmacol. Exp. Therap. 177, 155-167.

Youdim, M.B.N., Collins, G.G.S., Sandler, M., Bevan Jones, J.B., Pare, C.M.B. and Nicholson, W.J. 1972. Human brain monoamine oxidase: Multiple forms and selective inhibitors. Nature, in press.

Yuwiler, A. 1972. Stress. in Handbook of Neurochemistry, ed. A. Yuwiler, Plenum Press, New York, vol. 6, p. 103.

THE NEUROBIOLOGY OF MOOD AND PSYCHOSES

Morris A. Lipton, Ph.D., M.D.

Department of Psychiatry and Biological Sciences Research

Center, Child Development Institute, The University of

North Carolina, Chapel Hill, North Carolina

On several occasions I have had the privelege of having Dr. Kety discuss my presentations. This is the first time I have had the opportunity to discuss his, and I am flattered. I am also pleased to discuss the papers of Drs. Mandell and Stein. In doing so, I shall try to function as a critic and a clinician – Not in a destructive sense, I hope, but rather for the purpose of showing how incomplete our hypotheses still are and how they have trouble accommodating some important existing information about pharmacology and behavior at both animal and human level. Perhaps this will lead to methods for extending the information and refining the hypotheses.

The three speakers have addressed themselves to what we may for convenience call the central sympathetic nervous system. Dr. Mandell presented his work on the dynamics of biogenic amine neurotransmitter turonover in the brain. About five years ago I was dissatisfied with the product-feedback control mechanism for regulating norepinephrine (NE) biosyntheses and attempted to alter levels of tyrosine hydroxylase by stress and other means. I was unsuccessful and concluded incorrectly that this rate limiting enzyme was immutable. Since then Axelrod and Weiner have demonstrated that synthesis of enzyme is responsive to presynaptic neural activation. Now Mandell has evidence for both increases and decreases in the rate of synthesis and activity of this enzyme in response to drugs. These changes may involve allosteric alterations in the enzyme and alterations in sub-cellular membrane binding. It appears that in addition to the well established mechanisms of enzyme regulation, there may be some new ones for the control of biosynthesis of neurotransmitters in the brain.

149

Dr. Kety has offered a scholarly summation of the anatomy of
the central sympathetic system and of the evidence regarding its
role in adaptive behavior like arousal, the regulation of hunger
and thirst, aggression and responses to stress. He also touched
briefly on the role of this system in the pathological mood disorders of man, submitting some of the evidence for the "catecholamine
hypothesis" of mood disorders.

Dr. Stein presented an imaginative theory of schizophrenia in
man based upon a postulated genetic defect in brain dopamine beta
hydroxylase which he feels might lead to production of 6 hydroxydopamine (6OHDA) and consequent gradual destruction of the reward
systems.

Each speaker has expressed appropriate caution in presenting
and interpreting his data and yet the effective and enthusiastic
presentations may have tended to make the audience minimize these
cautions. I should like to amplify them.

For example, we may begin by asking whether the central sympathetic system is essential to life in the sense that the thyroid
gland or adrenal cortex is, or whether it is essential for those
evolutionary refinements of life which we call adaption, learning,
socialization or even pleasure.

That same question was asked many years ago by Cannon of the
adrenal medulla and the peripheral sympathetic system. You will
recall that removal of the adrenal medulla did not cause death nor
even gross deficits in adaptive behavior. Later it was found that
the failure to obtain life and death consequences was due to the
presence of the sympathetic nervous system with all of its available NE which could apparently substitute for the gland. If the
experiment is reversed and the peripheral sympathetic system is
destroyed surgically, immunologically or pharmacologically the
results are not devastating, presumably because the adrenal medulla
can compensate.

Suppose they are both destroyed; the adrenal medulla surgically
and the sympathetic nervous system in some other way? I have not
done an exhaustive literature survey, but I am under the impression
that this experiment has seldom been done. It seems to me that
Brodie gave bretyllium to an animal that had been adrenal demedul-
lated. Bretyllium blocks sympathetic transmission and in this sense
performs a chemical sympathectomy. The resulting animals were very
fragile, could not tolerate cold stress and died easily. So it
would certainly appear that the combination of the peripheral sympathetic nervous system and the adrenal medulla are together required for environmental coping and adaptation, perhaps for life
itself except in the most sheltered and stable environment.

What about the functions of the central sympathetic nervous
system? Dr. Kety summarized the pharmacological evidence for a
role of this system in a variety of adaptive processes. But a more
direct test of whether it is essential for life and what its functions
are could be obtained if it were possible to destroy this system
selectively, completely and irreversibly. This end is very nearly
achieved by administering (6-OHDA) into ventricles or cistern of
the brain. As Dr. Stein pointed out, this compound has the unique
property of being taken up by neurons containing catecholamines and
destroying them.

In the research laboratories of our Department of Psychiatry
and our Child Development Center, Drs. Breese, Grant, and Howard
are employing this method (Breese and Traylor, 1970a, 1970b). They
have been able to reduce the concentration of NE and dopamine (DA)
in the brain of rats by 90% (Breese and Traylor, 1970b). Tyrosine
hydroxylase also disappears and both the enzyme and the neurohumors
remain at very low levels for up to 80 days (Breese and Traylor,
1970a). Serotonin is not affected. An almost complete chemical
sympathectomy is therefore achieved. This is quite different from
the electrolytic lesions which destroy an anatomical area. Here,
so far as we can tell only catecholamine neurons are destroyed.
The fact that there is not quite 100% destruction somewhat compli-
cates the interpretation of results, because there may be great
redundancy in this system, or perhaps because crucial areas are not
affected. Yet it is worth noting the consequences on the vitality
and behavior of animals treated in this fashion.

Young adult rats prepared in this manner are surprisingly
robust. In appearance they are vastly different from the reserpinized rat (Breese and Traylor, 1970a). Following a transient
hypothermia lasting a few hours, the animals recover their temperature and are able to tolerate 24 hours in a cold room as well as
controls (Breese et al., 1972). If observed without disturbance in
their cages they cannot be distinguished from normal rats, although
there is a suspicion that they are less well groomed (Breese and
Traylor, 1970a). Activity in the cage or in an open field seems
normal. Males copulate and females have normal estrus cycles and
bear normal litters which they rear in a normal fashion (Grant et
al., 1971). Muricide as a test of aggressive behavior is not
different. In short, they are able to tolerate the ordinary stresses
of the laboratory rat's life. Certainly then the central sympathetic
system, with the caveats previously mentioned, does not seem to be
essential for life.

More complex tests do, however, reveal some differences. In
confirmation of the work of Stein, self stimulation in the reward
systems is markedly diminished (Breese et al., 1971). When placed
in a T maze and given an appetitive task, rats show a lower rate of

acquisition but ultimately the same level of performance as controls (Howard et al., 1971). When removed from the T maze task for a period there is a more rapid decrement in performance than in the controls. Interestingly, the performance in the T maze of 6-OHDA treated rats is exquisitely sensitive to alpha-methyl tyrosine (AMT) (Cooper et al, in press, a). Compared to controls such animals completely stop exploratory activity with small doses of AMT and appear much like reserpinized rats. Interpretation of these results is difficult, but if AMT can be assumed to do nothing but inhibit tyrosine hydroxylase then it would appear that the AMT may reduce the already depleted catecholamine levels below a crucial level and to the point where animals are immobilized. Still another interesting pharmacological interaction occurs when 6-OHDA treated animals are given parachlorphenylalanine (PCPA). Such animals are much more aggressive and vicious than PCPA treated controls. They attack the glove of the investigator, and show a marked increase in muricide. Again, it would appear that the 10% of catecholamines remaining after 6-OHDA treatment is crucial for the control of aggression.

Another deficit shows up when the animals are tested in an avoidance task. When placed in a two chambered shuttle box in which escape from foot shock is possible by an appropriate move to the opposite box, the animals show a major persistent deficit in avoidance behavior compared to controls (Cooper et al., in press, b).

In still another test comparing controls and 6-OHDA rats that have learned to press a bar for a food reward, it has been found that if the spring tension on the bar is increased so that more effort is required to press the bar, the 6-OHDA treated animals stop performing. Since the rats are the same size and weight it does not seem likely that they are weaker. Rather it seems that they are less motivated or that they may have an undetected motor defect - perhaps like in human Parkinson disease - which appears only under difficult work loads.

The experiments described thus far were all performed with young adult rats. If rats younger than seven days are given the 6-OHDA intracisternally, and are then observed over the next several months they show the same performance defects as the treated young adults, but to a significantly greater degree. In addition, they show some inhibition of growth which is not restored by bovine growth hormone (Breese and Traylor, 1972; Breese et al., in press).

It is undoubtedly premature to interpret the reasons for the performance deficits. In some cases the data would fit with decreased motivation as a cause. In other cases we would like to think that there is a decrement in learning ability and that the catecholamines are involved in acquisition and consolidation of

information. It would be very nice indeed if we have produced a
mentally retarded rat and that a central catecholamine deficit may
underlie some forms of mental retardation. But much work remains
to be done to test these hypotheses. Not the least of the jobs
facing our investigators is the tedious task of attempting to
replicate their work under conditions where NE alone, rather than
NE and DA are depleted together. Still another task is that of
further investigating the interaction of the serotonin and catecholamine systems.

Rats have a limited repertoire of spontaneous behavior, and
it is quite possible that we have not yet learned how to look at
them critically. It may therefore be of interest to you to have
me report on some very recent work on monkeys done by Drs. Prange,
Breese, Grant and Howard of our laboratories in collaboration with
Dr. McKinney who is working with Harlow's group at the University
of Wisconsin. In this study we do the chemistry and enzymology and
they do the behavior. The results are quite preliminary.

The macaque is quite sensitive to intraventricular 6-OHDA and
may convulse and die if a full dose is initially administered.
Probably this is because of the massive release of stored catecholamines following the initial injection. Consequently, animals are
given weekly injections of 2, 4, 8, and 16 mg of 6-OHDA progressively. This results in a level of brain NE which is 40-60% of normal,
but produces only minimal changes in DA. Following each injection
the animals show a syndrome which the investigators have termed
"amotivational" for about six hours. The animals are conscious
but relatively inactive and fail to respond to the social milieu of
the other monkeys. After that they recover and rejoin the activities of their group. On gross observation it is impossible to distinguish the behavior of these animals from that of controls. Perhaps more systematic studies will reveal differences and these are
in progress.

The monkey studies are exciting for many reasons. There is no
question but that the production of animal models of human illness
would help us considerably in the study of the human conditions.
There is some evidence that such models can be produced in primates
by psychological techniques and perhaps by pharmacological means as
well. In the study of human illness we are necessarily limited by
ethical and humane considerations. Thus, it has meant that our
experimental methods are limited to the chemical study of blood,
urine, and occasionally spinal fluid of patients, the electrophysiological study of the brain almost invariably with surface electrodes, and the retrospective correlation of the behavioral effects
of psychotropic drugs with their chemical and pharmacological action
in man and animals. Kety has summarized the chemical and pharmacological evidence that implicates the catecholamines in the mood

disorders of man. It is impressive evidence and leads me to feel
that what is currently known is correct but incomplete. For example, it is impressive that 15% of patients receiving reserpine for
hypertension become depressed even though they receive sufficient
reserpine for the treatment of their hypertension. Similarly,
patients on AMT rarely become clinically depressed although the
evidence for significant inhibition of NE synthesis is good. Urinary studies with 3-methoxy 4-hydroxy phenylglycol (MHPG), suppo-
sedly a relatively specific indicator of NE metabolism in the brain
as contrasted with the periphery, suggests that this compound is
lowered in depression and rises with clinical recovery, but again
the differences are small. The monkey studies which produce large
differences in the NE content of the brain will permit us to determine whether urinary MHPG truly reflects brain events. The behavioral studies will also permit us to further examine the function
of this system on mood and social interaction in species more
closely related to man. From such studies we may yet learn about
what must be added to our information about the role of NE to com-
prehend the neurobiology of depression.

    In summary, then, the problem of the role of the central sym-
pathetic nervous system is in that very exciting state where information is still meager but is accumulating very rapidly. Apparent
contradictions will undoubtedly be resolved, but at this moment
we may either be impressed with the evidence presented by Dr. Kety
which implicates the system in rage, hunger, thirst, stress and
arousal in animals and depression in man, or we may be equally
impressed by how little removal of 90% of the system does. Redundancy may be the answer. Perhaps the 10% remains in the crucial
areas. Perhaps other systems can compensate. Many other guesses
may be made, but it must be confessed that at this time we simply
do not know.

    May I now speak as a clinical psychiatrist and comment on Dr.
Stein's theory of schizophrenia. He postulates a genetic defect
in dopamine beta hydroxylase which results in a relative excess of
DA over NE in the catecholamine containing neurons of the reward
system. He further postulates that upon discharge of the amine
following stimulation, some of the DA is oxidized to 6-OHDA which
is taken up again into the neuron and gradually destroys it. The
consequence is a gradual reduction of the reward system and a
chronic state of anhedonia, which in Stein's view results in the
chronic chaotic thinking and lack of goal directed behavior which
Stein feels are characteristic of schizophrenia.

    Formally, a theory is an hypothesis with a high degree of
probability. What Dr. Stein calls a theory is therefore more
properly an hypothesis. It has little direct evidence to sustain
it. It is elegant and so simple that those of us who have strug-

gled with attempts to understand the mechanisms of the schizophrenic cannot help but be envious and critical.

Dr. Stein's concept of schizophrenia is based on concepts
originally voiced by the psychoanalysts Rado and Sullivan who
emphasized that schizophrenics are devoid of the hope and perhaps
capacity for pleasure and instead seek only security. Meehl in a
recent Presidential address to the American Psychological Association also considers this possibility (among others) as characteristic of a schizophrenic genotype.

But there are other characteristics of schizophrenia which cannot be ignored. Prominent among these is the thought disorder or
what has been termed cognitive slippage. There is the marked ambi-
valence which can be looked at as the consequence of an unresolved
intense approach avoidance conflict. There is the social aversiveness, and finally there are the delusions, hallucinations and other
secondary symptoms. It is possible that all of these are consequences of a primary anhedonia, but there are other possibilities
as well. For example, the learning of associations may be due to
a state of hyperarousal associated with a loss of inhibitory circuits. We are all capable of loose associations in dreams and day
dreams and on the psychoanalyst's couch where we consciously suspend all critical inhibitions of our thought processes. But we are
able to inhibit and organize our thinking at will by selective attention. The schizophrenic seems unable to do this. Similarly,
the cognitive slippage may result from the inability to retain a
short term memory trace long enough to make a set, as Dr. Callaway
suggested.

All of these possibilities could be genetically based and to
be wary of Dr. Stein's formulation is not to deny a genetic dia-
thesis. We must constantly remind ourselves that there are probably many schizophrenics and that he may be describing a subtype.
The best genetic data suggests the involvement of more than one
gene. We must also remind ourselves that twin studies, though
focussing on the genetic component still leave a majority of the
influence in the development of the clinical condition or phenotype
in the hands of the environmentalists. Once again, ethical constraints prohibit us from anything except natural experiments in
man. Here too, primate studies such as Harlow has done involving
psychological manipulation which produce devastating behavioral
effects may offer a useful tool to the biologists capable of
investigating brain function. For example, it would be fascinating
to learn whether such disturbed monkeys would still seek self stimulation if electrodes were implanted in their reward systems.

Stein's hypothesis does not fit the clinical data completely.
For example, token economies based upon reward are successfully

used with chronic schizophrenics. Although the schizophrenic's pleasure in interpersonal situations is markedly diminished and is associated with their social aversiveness, they seem capable of deriving pleasure from impersonal objects. Pharmacologically, Stein argues that chlorpromazine helps the schizophrenic by inhibiting the reuptake of the 6-OHDA, thus protecting the reward system. If this were the mode of action, we might expect imipramine to do as well, but imipramine is clinically effective only in the depressed schizophrenic, altering his mood but not his thought processes. Furthermore, chlorpromazine, like so many of our drugs has more than one mode of action. As Dr. Kety pointed out it also blocks catecholamine receptors and may thus decrease the hypervigilance.

Fortunately, Dr. Stein's hypothesis is in some ways testable, the preliminary data he has furnished which show both low and high dopamine beta hydroxylase are exciting and we can look forward to additional data. Perhaps there will be two populations of schizophrenics. Indeed, the capacity to test his hypothesis in man already exists, for if I recall correctly, Heath at Tulane has schizophrenic patients with electrodes already implanted in their reward centers and may already have done this study.

Clearly, concentrations of biogenic amines at synaptic clefts must be considered in relation to the sensitivity of the receptors and effectors upon which they act. This is one of the directions in which research at the University of North Carolina has moved and we have obtained some gratifying clinical and experimental results by studying the interaction of thyroid hormone, which we believe sensitizes receptors, and antidepressant drugs. Clearly too, the catecholamine system does not work alone, but in conjunction with the serotonergic and cholinergic systems. Their interaction must be unravelled. Finally, in man at least, we have the phenomenon that we are able to stimulate ourselves and to generate all types of effects through our memories. How this is achieved and what might be done to alter them remains another task for the future.

REFERENCES

Breese, G.R., Howard, J.L. and Leahy, J.P. 1971. Effect of 6-hydroxydopamine in electrical self-stimulation of brain. Brit. J. Pharmacol., 43, 255-257.

Breese, G.R., Moore, R. and Howard, J.L. 1972. Central actions of 6-hydroxydopamine and other phenylethylamine derivatives in body temperature in the rat. J. Pharmacol. Exp. Ther., 180, 591-602.

Breese, G.R., Smith, R.D., Cooper, B.R., Howard, J.L. and Grant, L.D. Biochemical and behavioral changes induced by 6-hydroxydopamine (6-OHDA) in the developing rat. Fifth Int. Congress on Pharmacol. In press.

Breese, G.R. and Traylor, T.D. 1970a. Effect of 6-hydroxydopamine
  on brain norepinephrine and dopamine: Evidence for selective
  degeneration of catecholamine neurons. J. Pharmacol. Exp. Ther.
  174, 413-420.
Breese, G.R. and Traylor, T.D. 1970b. Depletion of brain noradrenaline and dopamine by 6-hydroxydopamine. Brit. J. Pharmacol.,
  42, 88-99.
Breese, G.R. and Traylor, T.D. 1972. Developmental characteristics
  of brain catecholamines and tyrosine hydroxylase in the rat:
  Effects of 6-hydroxydopamine. Brit. J. Pharmacol., 44, 210-222.
Cooper, B.R., Breese, G.R., Howard, J.L. and Grant, L.D. Enhanced
  behavioral depressant effects of reserpine and alphamethyl
  tyrosine after 6-hydroxydopamine. Psychopharmacol. (Submitted
  for publication, a).
Cooper, B.R., Breese, G.R., Howard, J.L. and Grant, L.D. Effect of
  central catecholamine alterations by 6-hydroxydopamine on shuttle
  box avoidance acquisition. Physiol. Behav. (Submitted for
  publication, b).
Grant, L.D., Sar, M., Stumpf, W.E., Howard, J.L. and Breese, G.R.
  1971. Effects of 6-hydroxydopamine (6-OHDA) on reproductive
  functions. Pharmacologist, 13, 286.
Howard, J.L., Grant, L.D. and Breese, G.R. 1971. Influence of 6-
  hydroxydopamine (6-OHDA) on performance in a modified T-maze.
  Pharmacologist, 13, 232.
McKinney, W., Breese, C.R., Howard, J.L. and Prange, A.J., Jr.
  Unpublished results.

THE EFFECT OF SHORT EXPERIENCES ON THE INCORPORATION OF RADIOACTIVE

PHOSPHATE INTO ACID-EXTRACTABLE NUCLEAR PROTEINS OF RAT BRAIN[1,2]

Edward Glassman, Barry Machlus and John Eric Wilson

The Neurobiology Program and The Division of Chemical
Neurobiology of the Department of Biochemistry, School
of Medicine, The University of North Carolina, Chapel
Hill, North Carolina 27514

It is generally accepted that the associative processes that
go on in the brain during learning involve the formation of functionally new neuronal pathways or networks, the nature of which
encode the memory. It is not known how the pathways or networks
are selected, or what the relationship of the pathway or network
is to the encoded information. This process of changing interneuronal communication and information flow in the brain must involve
molecular changes that produce new properties in these neurons. It
should be clear from the onset that the molecules themselves do not
encode memory within their chemical structures, but act only to
affect the neurons so new pathways and networks of interneuronal
communication can form. Many hypotheses have been proposed concerning these molecular changes. In general these involve chemical
changes that affect the efficiency of synaptic transmission or the
connectivity between neurons. This could happen, for example, if
there were changes in the amount of transmitter released, changes
in the rate of transmitter destruction, changes in the size of the
synapse, or alterations in the number or activity of receptor sites
for the transmitter. Thus the possible number of theoretical models
is very large. The major problem is the generation of data that
will have bearing on the molecular events associated with changes
in neuronal connectivity.

Our approach is based on the research that suggests that the
storage or consolidation of memory involves at least two successive
stages, short and long term memory. The ability to learn or acquire
short term memory does not seem to be dependent on high levels of
RNA or protein synthesis. This notion is based partly on the report
that extensive neurophysiological activity and behavioral phenomena

159

(short term habituation) can take place in Aplysia even though protein synthesis is inhibited over 95% (Schwartz et al., 1971), and
on reports that extensive inhibition of RNA or protein synthesis in
the brains of goldfish (Agranoff, 1968) or mice (Barondes and Cohen,
1967; Flexner and Flexner, 1968) that does affect the formation of
long term memory, does not interfere with the primary acquisition
of tasks involving conditioned avoidance. One must conclude that
if RNA or protein is playing a role in short term changes in neuronal connectivity, the mechanism more likely involves conformational
changes in pre-existing molecules rather than changes in synthesis.
The possibility of a requirement for very low levels of synthesis,
or the synthesis of other molecules is not ruled out except by the
short time parameters involved.

Long term memory deficits are produced, however, when inhibitors of protein or RNA synthesis are given before or immediately
after training. This suggests that the formation of long term memory
may depend on the synthesis of these macromolecules. The perpetua-
tion of long term memory does not seem to be specifically dependent
on the continuous synthesis of RNA and protein, since long term
memory is not affected by these inhibitors once it has been established.

If the formation of long term memory involves high levels of
synthesis of RNA and protein, then it would seem possible to detect
this synthesis through the use of radioactive tracers. A behavioral
task that is rapidly learned was used in order to measure the incorporation of precursors into RNA, proteins, lipids, and other substances during the time long term memory is supposed to be forming.
It was hoped that in this way the synthesis of molecules associated
with the formation of long term memory might be detected.

It was shown utilizing a double isotope technique that mice
injected with radioactive uridine and then trained for 15 minutes
to avoid a shock by jumping to a shelf incorporated about 40 percent
more radioactivity into brain RNA and into polysomes than did untrained mice (Zemp et al., 1966; Adair et al., 1968a). No differences were found in liver or in kidney. Localization by various methods has shown that the bulk of the increased incorporation of uridine
into RNA of the trained mouse is in structures in the core brain
(Zemp et al., 1967; Kahan et al., 1970). Lesser changes occur
elsewhere.

Sucrose density gradient centrifugation showed that the increased radioactivity associated with the RNA of the trained mouse was
heterogeneous with respect to sedimentation rate, and a unique species of radioactive RNA was not detectable by this means (Zemp et
al., 1966). The patterns of radioactivity were of similar shape for
both trained and untrained mice, and resembled those found after RNA

synthesis has been stimulated in other tissues by hormones. This
suggests that the increased radioactivity is not confined to a
single species of RNA, and that a generalized metabolic stimulation
of relevant brain cells had occurred. In addition, the increased
incorporation into polysomes of the brain during the training experience suggested that the increased radioactivity is either in messenger RNA or in preribosomal RNA (Adair et al., 1968a; Coleman et
al., 1971). Thus, the brain responds biochemically in a manner
similar to chemical responses found in other tissues. The specificity seems to be in the stimulus that each tissue or cell responds
to. There was no evidence for a unique brain RNA with a function
that does not involve protein synthesis; indeed the brain RNA was
similar to RNA extracted from other tissues.

Because of their hypothesized relationship to RNA synthesis,
the effect of a short training experience on the phosphorylation of
acid-soluble proteins in the nuclei of rat brain cells has also been
examined. This paper is a summary of part of that research (Machlus,
1971). Preliminary reports of this work have appeared elsewhere
(Machlus et al., 1971a, 1971b; Glassman and Wilson, 1972).

The training apparatus was a modification of that described by
Coleman et al. (1971b). It consisted of a runway, the starting end
of which was 40 cm long, 56 cm high and 15 cm wide, with black walls
and a shock grid floor made of 3/32 in brass rods, the centers of
which were 8.5 mm apart. The escape area was 26 cm long, 48 cm high
and 15 cm wide. It was elevated 8 cm above the grid floor and had
white walls.

The rats were gentled by handling them for 5 minutes on each of
two days prior to the experiment. Before training they were given
5 minutes to explore the escape platform and the grid floor. The
rat was held by the tail and lowered onto the start area of the grid
floor. A switch on a Lafayette timer (model 5001A) was turned on as
the animal was released. Five seconds later, a foot shock of 0.8 mA
was activated (Lafayette master shocker with scrambler and inter-
rupter model A615C) and remained on for 25 seconds. If the rat
stepped back onto the grid, it received footshocks until it remounted
the platform. The trial lasted 30 seconds. At the end of this time
the rat was picked up by the tail and the next trial started. Training continued for 5 minutes and consisted of 10 trials. Animals were
judged to have made an avoidance response if they ran from the grid
floor to the platform before the onset of the shock. The average
number of avoidances was 6 out of the 10 trials. Learning curves
and further details are shown in Coleman et al. (1971a), who reported
that 15 to 20 minutes of this training procedure produces increased
incorporation of radioactive uridine into polysomes of rat brain.

A double isotope method was used throughout. A pair of male

Wistar albino rats approximately 100 days old obtained from Research
Animals (Braddock, Pennsylvania) was injected under light ether
anesthesia through the eye socket into the brain. This was accomplished by pushing aside the lacrimal duct and passing the needle
through the optic foramen. Inspection revealed that the needle
point penetrated the basal forebrain and it is believed that this is
the area in which the injected material is released. One rat of the
pair received 0.1 mCi of $H_3{}^{32}PO_4$ (1.1 x 10^{-8} mmoles of phosphate
from Tracerlab) in 100 μl of 0.02 $\underline{N}$ HCl, while the other rat received
0.1 mCi of $H_3{}^{33}PO_4$ (1.9 x 10^{-8} mmoles of phosphate from Tracerlab)
in 100 μl of 0.02 $\underline{N}$ HCl. It is of interest that there is approximately 100 μmoles of phosphorus per rat brain (Davison and Dobbing,
1970). Since a single injection contained about 10^{-8} mmole of phosphate, the phosphorus added by the injection was about 10^{-7}% of the
total present. The rats were coded and returned to their cages.
After 25 minutes, one rat was placed in the training apparatus for
5 minutes of adaptation, after which it was trained for 5 minutes.
The rat was then immediately sacrificed by decapitation and its brain
was homogenized gently and kept in ice until the brain of the other
rat was ready for homogenization, usually about 10 minutes later.
The other rat was not trained. It was sometimes placed in the training apparatus for 10 minutes 25 minutes after injection, but usually
it was kept quiet for 35 minutes. The brain of this rat was then
immediately removed and homogenized in the same homogenizing tube
containing the homogenized brain of the trained rat. The order in
which the trained and untrained rats were sacrificed was randomized.

Fractions containing nuclei were isolated and extracted with
0.2 $\underline{N}$ HCl. The soluble proteins were precipitated with acetone,
dried with ethanol and ether, and then dissolved in 1 ml of 7% guanidinium chloride. About 250 γ of protein were applied to a 0.9 x
15 cm Amberlite IRC-50 column, and eluted first with 100 ml of a
linear gradient of guanidinium chloride from 7% to 14%, and then
with 25 ml of 40% guanidinium chloride in 0.1 $\underline{M}$ potassium phosphate,
pH 6.8. One ml samples were collected. The flow rate was one drop
every 30 seconds, and protein recovery was over 95%. Protein was
determined turbidimetrically at 400 mμ after adding TCA, and radioactivity was determined in a scintillation counter.

To correct for experimental errors in the amounts of radioactive phosphate injected and distributed throughout each brain, the
ratio of ^{32}P to ^{33}P in AMP isolated from the original homogenate was
used as a correction factor. The ratio of ^{32}P to ^{33}P in GMP, CMP
and UMP of the trained and untrained rats was similar to that in AMP.

There are four main protein peaks following column chromatography of the total acid-soluble nuclear proteins from brain on
Amberlite IRC-50 resin. These correspond to, in order of elution,
non-histone acid-soluble protein (NAP), which are not retarded by

the column under the conditions used, the lysine rich histones, the slightly lysine rich histones, and the arginine rich histones.

Preliminary experiments indicated that the training experience affected the amount of radioactivity in NAP and histones. Time course experiments showed that there is about twice the amount of radioactive phosphate in NAP from the trained rat's brain at 5 minutes of training, but by 10 minutes there is no significant difference between the trained and untrained rats. These results indicate the effect of the behavior on the phosphorylation of NAP is short-lived, and suggest that this chemical response might be a trigger for other later occurring reactions.

We now have data comparing the amount of radioactive phosphate in NAP in 30 trained and 30 untrained rats sacrificed immediately after the 5 minute training period. These experiments were conducted blind, and were carried out over a period of months. Although there was no significant difference in the average amount of radioactivity in the AMP from the brains of trained and untrained rats, the trained rats showed an average of 109 ± 18% (S.D.) more radioactivity in brain NAP than untrained rats. The increase in radioactivity is not due to an increase in amount of brain NAP, since the amount of brain NAP does not significantly change as a result of training. On the other hand, comparison of 5 trained and 5 untrained rats showed an average of 59 ± 40% (S.D.) less radioactivity in brain histones of the trained rats.

Additional fractionation on an Amberlite IRC-50 column and by electrophoresis on polyacrylamide gels demonstrated that NAP is a heterogeneous mixture of proteins, not all of which are phosphorylated. The chromatography was done on a column of Amberlite IRC-50 resin eluted with a 0 to 7% linear gradient of guanidinium chloride in 0.1 M potassium phosphate, pH 6.8. The results indicated two to three distinct peaks, only one of which showed radioactivity. The result of disc electrophoresis showed that at least fourteen different bands of proteins were present in NAP, but only three bands were radioactive. It is of interest that in all three bands, the proteins isolated from the brain of the trained rat showed a greater amount of radioactivity than the protein from the brain of the untrained rat. These proteins are now being purified further.

The possibility remained that the phosphate was not incorporated into NAP, but into contaminating RNA or phospholipids. In order to test this, the NAP containing ^{32}P and ^{33}P from trained and untrained rat brains was subjected to treatment with various enzymes. The data showed that RNase, DNase, and phospholipase-C were ineffective in removing the radioactive phosphate from NAP. Only proteolytic enzymes had any effect; as expected pronase was the most effective of these enzymes.

To examine this further, NAP prepared separately from trained and untrained rats was hydrolyzed with pronase, and amino acid analysis carried out. The increase in radioactive phosphate in brain NAP from the trained rat was accountable for as an increase in radioactivity in phosphoserine. It is of interest that there was no difference in the amount of radioactive phosphate in phosphothreonine in brain NAP between trained and untrained rats, although phosphothreonine contained large amounts of radioactivity.

When compared separately, the molar ratio of phosphothreonine to threonine in NAP is approximately the same in both trained and untrained rats, whereas the ratio of the amount of phosphoserine to serine in NAP is 2.58 in the trained rats and 0.95 in the untrained ones. This finding of an increase in the amount of phosphoserine relative to serine in the brain NAP of the trained rat is consistent with the increase in radioactive phosphate. Of interest, this increase in amount of phosphoserine in brain NAP of the trained rat could also be observed in the absence of any injection of radioactive phosphate; these data eliminate the injection as playing any role in these phenomena.

This difference between the amounts of phosphoserine from brain NAP from trained and untrained rats may be due to increased phosphorylation or decreased dephosphorylation of NAP in the trained animal, or to any one of a number of other alternatives. It is not possible at present to shed light on this question because of technical difficulties and thus we refer only to the difference in the amount of radioactive phosphate, and do not specify a mechanism of how it comes about. However, it is improbable that the increase in radioactive phosphate in brain NAP from the trained rat is due to a change in permeability of brain cells to radioactive phosphate; nor is it likely to be due to changes in the phosphate or nucleotide pools, since it is difficult to explain the simultaneous lack of difference in the amount of radioactive phosphate in phosphothreonine in trained and untrained rats, and the decrease in the amount of radioactive phosphate in histones on this basis. Indeed these data also rule out a change in the rate of production of the phosphate donor (ATP?) to the nuclear protein unless one postulates different specificity in the reactions involving the phosphorylation of histones, and the serine and threonine in NAP.

A number of important questions remain to be answered. The areas of the brain, and even more specifically, the exact cells which are responding to this stimulus are not known. In preliminary experiments the area of the brain involved in this chemical response was localized by grossly dissecting the brain into 4 parts. Only the lower half of the cerebrum, the part containing the amygdala, entorhinal cortex, hypothalamus, mid-brain tegmentum, posterior-ventral hippocampus and other structures showed more radioactive phosphate in brain NAP of the trained rat. More work is necessary for more precise localization, possibly by autoradiography. It is of interest that this is also the region involved in the increased incorporation of radioactive uridine into RNA of mice undergoing jump box training as shown in previous studies (Kahan et al., 1970; Zemp et al., 1967).

Another important problem is that the behavioral or environmental agents that caused such chemical changes in the brain of the trained rat have not been elucidated. It may be that the learning, per se, or the special stresses and emotional and motivational effects of learning are responsible for triggering such chemical responses. We have performed experiments on rats lacking the pituitary or adrenal glands and showed that these glands are not necessary for the increase of radioactive phosphate in brain NAP.

It is also possible, however, that these chemical changes are due to non-specific stimuli, or to the activity associated with the training experience. Visual (Appel et al., 1967; White and Sundeen, 1967; Talwar et al., 1966), auditory (Hamberger and Hyden, 1945), rotary (Hamberger and Hyden, 1949a, 1949b; Watson, 1965; Attardi, 1957; Jarlstedt, 1966a, 1966b), olfactory (Rappoport and Daginawala, 1968), and stress stimulation (Bryan et al., 1967; Altman and Das, 1966) have been reported to cause changes in RNA or polysomes in the nervous system, and such stimuli could be effective during training. To determine whether these differences were due to the stimulation the trained animal received from the handling and shocks or to its activity, the effects of two additional behaviors on this chemical response were compared with quiet rats; these were animals who were denied access to the safe area, but who received random shocks, or who received shocks paired with the training record of a previously trained rat.

There were no differences between the rats undergoing these behaviors and quiet rats with respect to the amount of radioactivity incorporated into brain NAP. These results clearly show that the environmental stimulation that the trained rat receives is not the cause of the increased radioactive phosphate in brain NAP in the trained rats.

To test whether the performance or recall of the task had some relevance, rats were trained for 5 minutes on each of 6 successive days. On the seventh day, they were injected with the radioactive phosphate using the standard double isotope method and allowed to perform in the runway for 5 minutes, after which they were sacrificed. Their brains were homogenized with brains of naive quiet rats.

The results were unexpected, since prior trained performing rats showed about the same increase in radioactive phosphate as do

naive rats when they are first trained. It became necessary to
determine the minimum amount of reminding necessary to increase the
amount of radioactive phosphate in brain NAP from the prior trained
rat. The data showed that placing a prior trained rat on the safe
platform for 5 minutes could elicit this increase in radioactive
phosphate in brain NAP, as well as merely injecting the prior
trained rat and placing it back in its own cage.

It is of interest that rats that were merely shocked at random
for 5 minutes per day for 6 days did not show an increase of radioactive phosphate in brain NAP when shocked or kept quiet on the
seventh day.

The results are interesting since we are comparing three groups
of rats whose treatment is identical, i.e., each is picked up,
injected with radioactive phosphate, placed in a quiet cage and
sacrificed 30 to 35 minutes later. The differences between the
groups lie in their past histories; one group is naive, one group has
been trained daily for 5 minutes for 6 days, and one group has been
given random shocks daily for 5 minutes for six days. Only the
prior trained group shows the increase in radioactive phosphate in
brain NAP.

Because Adair et al. (1968b) reported that the performing prior
trained mouse does not have increased amounts of radioactive uridine
in polysomes, some of these experiments were repeated in the mouse
using the jump box (Zemp et al., 1966; Schlesinger and Wimer, 1967)
and the alleyway that was used for the rat. The biochemical procedure for the mouse was exactly the same as for the rat. The only
difference was that one mouse of a pair received 0.033 mCi of
$H_3{}^{32}PO_4$ (0.364 x 10^{-8} μmoles of phosphate from Tracerlab) in 33
μl of 0.02 N HCl, while the other mouse received 0.033 mCi of
$H_3{}^{33}PO_4$ (0.641 x 10^{-8} μmoles of phosphate from Tracerlab) in 33 μl
of 0.02 N HCl.

The data on mice are not extensive, but show that training in
the runway or the jump box caused increased amounts of radioactive
phosphate in brain NAP in the trained mouse. In addition, prior
trained mice performing in the jump box or the runway also showed
an increase of phosphate in brain NAP. Classical conditioning of
the mouse in the jump box for 5 minutes had no effect, however, a
result of great interest since Adair et al. (1968b) also showed no
effect of classical conditioning on the incorporation of uridine
into brain polysomes.

Thus the increase of radioactive phosphate in brain NAP does
not take place in yoked rats, in randomly shocked rats or in clas-
sically conditioned mice. These results eliminate the shocks, the
handling and the other stimulation the trained animal receives, as

well as his increased activity as the possible cause of the increase.
The increase of radioactive phosphate in brain NAP does take place,
however, if animals that were prior trained for 6 days are made to
perform the task for 5 minutes, placed on the safe platform for 5
minutes, or handled and then returned to their home cage. Even rats
that are prior-trained for only one day show more radioactive phosphate in brain NAP when reminded of the experience the next day.

It is therefore obvious that the discussion of the behavioral
trigger by Glassman and Wilson (1970), in which it was concluded
that the insight development phase exclusively contained the behavioral trigger for increased incorporation of uridine into brain RNA
does not apply here. It appears that learning this avoidance conditioning task has a permanent biochemical or physiological effect on
the rat or mouse so that increased incorporation of phosphate into
brain NAP is triggered by subtle reminders yet to be determined.
It may well be that the rate of turnover of phosphate in NAP is
permanently increased by the training experience, but the finding
that the effect is transient on the first day of training would
argue against this. It would be of great interest to determine whether the decrease in amount of radioactive phosphate in brain histones in the trained rat also occurs in a reminded prior-trained
animal.

It is tempting to speculate that a specific emotional response
generated by the training experience is actually the trigger, and
that this response can also be evoked when prior trained animals are
reminded of their experience at a later time. This would be consistent with the fact that the chemical response appears to take place
in the basal forebrain, an area of the brain thought to be involved
in emotional responses and arousal. The fact that a rat that is not
trained, but has been randomly shocked for 5 minutes per day for 6
days does not show an increase in radioactive phosphate in brain NAP
when reminded of this experience on the seventh day would suggest
that the learning aspect of the training experience is extremely
important to this phenomenon. In the absence of additional data, it
would be premature to speculate further on the significance of these
phenomena.

The chemical changes in the reminded animals have implications
for the data of Hyden and associates (Hyden and Lange, 1970), and
others, who train for many days, or even weeks, at the end of which
chemical analysis is carried out. It is not clear whether chemical
differences at this time reflect the effects of primary acquisition,
long term memory formation, a reminder effect, or none of these.
These results also have implications for the reported effects of
enriched environments (Rosenzweig et al., 1972). If repeated reminders of a training experience can elicit chemical synthesis, then a
series of reminders could perhaps account for the effects of

enriched environments, particularly since these effects are now
shown to be transitory.

Until it is known whether these changes in phosphorylation
have anything to do with the learning process per se, or with an
incidental process, it will be extremely difficult to correlate
our results with the formation of long term memory, although it is
extremely tempting to do so. The increased phosphorylation of NAP
might trigger the synthesis of RNA that codes for proteins involved
in this process. These protein(s) might cause the synaptic associations that developed during short term learning to become
permanent. The protein(s) may be related to the peptide(s) postulated to be involved in the maintenance and retrieval of memory by
Flexner and Flexner (1968) and Bohus and de Wied (1966). Alternatively, the increased phosphorylation of NAP may be a correlate
of recall. One interesting possibility deserving further investigation is that phosphorylation of proteins at the synapse or in
the cell may bring about conformational changes that affect connectivity and thus lead to short term memory. We are looking into
this possibility.

In this paper we have described the effect of a short training
experience on the turnover of radioactive phosphate in various
chemicals in rat brain. First, there is little or no effect on AMP,
CMP, UMP, or GMP. Second, the radioactivity is increased in NAP,
but decreased in histones. One purpose of this research is to
attempt to understand the significance of the reported effects of
behavior on macromolecules in the nervous system (see Glassman,
1969, for a review of this field). A detailed discussion of the
role of these macromolecules in the learning processes would be very
premature; there is no clear evidence that these macromolecules play
a direct role in encoding, and cause and effect relationships in this
area are still difficult to prove. The long-range problems that seem
important are the aspect of the environment or the behavior that
triggers these chemical responses, the physiological and biochemical
processes that convey the information from outside the animal to the
cells that are responding, and the significance of the chemical response in terms of the functional role that the macromolecules play
in the operations of these cells of the nervous system. Even if no
aspect of the learning process is involved in the changes reported
here, the fact that behavior can affect macromolecules should be of
great interest to those studying nervous system function. The role
these macromolecules play in the nervous system may be only the role
they play in the function of all tissues, i.e., to maintain cellular
health, and structure, and to enable the cellular machinery to function. If so, it is important to know this, and the solutions to the
problem of the unique functions of the nervous system in regulating
behavior can be sought elsewhere.

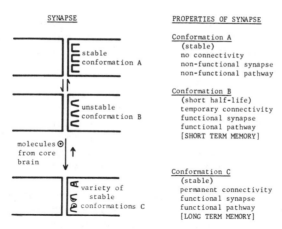

Figure 1. Model for a molecular basis for short and long term memory. See Postcript for an explanation.

POSTCRIPT

It is often useful to have a molecular model in mind when planning research of this type. Figure 1 depicts a model that has helped us in our thinking and we present it here mainly for its heuristic value, and not because we feel that such an oversimplified view of memory processes is adequate.

The model postulates that a chemical located at the synapse or elsewhere in the pre- or postsynaptic neuron is critical for connectivity, and that the control of connectivity is exerted by the conformational state of this chemical. Conformation state A is stable, but connectivity is not facilitated. If training occurs and the various associative processes that take place select a pathway or network in which this synapse is important, the molecule is shifted into conformational state B. There are several guesses as to how this occurs, but mere electrical activity probably does not produce this change. Conformational state B is unstable, but it does change the properties of the neuron so that a temporary change in connectivity occurs. The new pathway or network, for as long as it is operating, can be considered a part of the mechanism underlying short term memory. The model then postulates that the molecule in conformation state B can have one of two fates. The molecule can revert back to conformation A, in which case connectivity is again as it was, and the memory is lost, or is no longer easily retrievable. Alternatively, if certain physiological or emotional factors are stimulated in the animal undergoing training, parts of the core brain (basal forebrain) are stimulated to manufacture chemicals that can convert the molecule in conformation B to a new stable form, conformation C, so that connectivity is permanently changed. If this occurs, then long term memory now

exists. Since the molecules made in the core brain interact only
with conformation B, only those pathways or networks that have been
selected as providing adaptive behavior will be affected and be made
permanent.

This model implies that the pathways and networks that are
part of the underlying basis for short and long term memory are
the same, or at least overlap to a great extent. It also implies
that the steps in the synthesis or distribution of molecules made
in the core brain are the ones sensitive to many, if not all, of the
agents that affect the formation of long term memory.

Much work is necessary to determine the validity of ideas like
this, but at present the model does provide a molecular basis for
the formation of pathways that encode short and long term memory,
and however simplistic, does have testable aspects.

ACKNOWLEDGMENTS

[1]This research was supported by grants from the U.S. Public Health
Service (MH18136, NS07457), the U.S. National Science Foundation
(GB18551), and the Ciba-Geigy Corporation.

[2]In this paper, the abbreviation NAP refers to non-histone acid-
soluble protein isolated from brain nuclei.

REFERENCES

Adair, L.B., Wilson, J.E. and Glassman, E. 1968a. Brain function
    and macromolecules. IV. Uridine incorporation into polysomes of
    mouse brain during different behavioral experiences. Proc. Nat.
    Acad. Sci., 61, 917–922.
Adair, L.B., Wilson, J.E., Zemp, J.W. and Glassman, E. 1968b.
    Brain function and macromolecules. III. Uridine incorporation
    into polysomes of mouse brain during short-term avoidance condi-
    tioning. Proc. Nat. Acad. Sci., 61, 606–613.
Agranoff, B.W. 1968. Actinomycin-D blocks formation of memory of
    shock-avoidance in goldfish. Science, 158, 1600–1601.
Altman, J. and Das, G.D. 1966. Behavioral manipulations and pro-
    tein metabolism of the brain: Effects of motor exercise on the
    utilization of leucine-^{3}H. Physiol. Behav., 1, 105–108.
Appel, S.H., Davis, W. and Scott, S. 1967. Brain polysomes: Re-
    sponse to environmental stimulation. Science, 157, 836–838.
Attardi, G. 1957. Quantitative behavior of cytoplasmic RNA in rat
    Purkinje cells following prolonged physiological stimulation.
    Exp. Cell Res., 4, 25–53.
Barondes, S.H. and Cohen, H.D. 1967. Delayed and sustained effect
    of acetoxycycloheximide on memory in mice. Proc. Nat. Acad. Sci.,
    58, 157–164.

Bohus, G. and de Wied, D. 1966. Inhibitory and facilitatory effect
of two related peptides on extinction of avoidance behavior.
Science, 153, 318-320.
Bryan, R.N., Bliss, E.L. and Beck, E.C. 1967. Incorporation of
uridine-^{3}H into mouse brain RNA during stress. Federation Proc.,
26, 709.
Coleman, M.S., Wilson, J.E. and Glassman, E. 1971a. Brain function
and macromolecules. VII. Uridine incorporation into polysomes of
mouse brain during extinction. Nature, 229, 54-55.
Coleman, M.S., Pfingst, B., Wilson, J.E. and Glassman, E. 1971b.
Brain function and macromolecules. VIII. Uridine incorporation
into brain polysomes of hypophysectomized rats and ovariectomized
mice during avoidance conditioning. Brain Res., 26, 349-360.
Crestfield, A.M., Moore, S. and Stein, W.H. 1963. The preparation
and enzymatic hydrolysis of reduced and S-carboxy-methylated pro-
teins. J. of Biol. Chem., 238, 622-627.
Davison, A. and Dobbing, J. 1970. Applied Neurochemistry, Davis
and Co., Philadelphia.
Dounce, A.L. 1971. Nuclear gels and chromosomal structure.
American Scientist, 59, 74-83.
Ellingson, J.S. and Lands, W.E.M. 1968. Phospholipid reactivation
of plasmalogen metabolism. Lipids, 3, 111-120.
Flexner, L.B. and Flexner, J.B. 1968. Intracerebral saline:
Effect on memory of trained mice treated with puromycin. Science,
159, 330-331.
Gershey, E.L., Vidali, G. and Allfrey, V.G. 1968. Chemical studies
of histone acetylation. J. of Biol. Chem., 243, 5018-5022.
Glassman, E. and Wilson, J.E. 1970. The effect of short experiences
on macromolecules in the brain. In: Biochemistry of Brain and
Behavior, R.E. Bowman and S.P. Datta (Eds.), Plenum Press, New
York, pp. 279-299.
Glassman, E. and Wilson, J.E. 1972. Changes in macromolecules
associated with memory processes in mammals. In: Macromolecules
and Behavior, J. Gaito (Ed.), Appleton-Century-Croft, New York,
pp. 39-46.
Glassman, E. 1969. The biochemistry of learning: An evaluation
of the role of RNA and protein. Ann. Rev. Biochem., 38, 605-646.
Hamberger, C.A. and Hyden, H. 1945. Cytochemical changes in the
cochlear ganglion caused by acoustic stimulation and trauma. Acta
Oto-Laryngol., suppl., 61, 1-29.
Hamberger, C.A. and Hyden, H. 1949a. Transneuronal chemical changes
in Deiters' [sic] nucleus. Acta Oto-Laryngol., suppl., 75, 82-113.
Hamberger, C.A. and Hyden, H. 1949b. Production of nucleoproteins
in the vestibular ganglion. Acta Oto-Laryngol., suppl., 75, 53-81.
Hirs, C.H.W., Moore, S. and Stein, W.H. 1953. A chromatographic
investigation of pancreatic ribonuclease. J. of Biol. Chem., 200,
493-506.
Hyden, H. and Lange, P.W. 1970. Protein changes in nerve cells
related to learning and conditioning. Neurosciences Second Study

Program, F.O. Schmitt (Ed.), Rockefeller University Press, New York, pp. 278-289.

Jarlstedt, J. 1966a. Functional localization in the cerebellar cortex studied by quantitative determinations of Purkinje cell RNA. I. RNA changes in rat cerebellar Purkinje cells after proprio- and exteroreceptive and vestibular stimulation. Acta Physiol. Scand., 67, 243-252.

Jarlstedt, J. 1966b. Functional localization in the cerebellar cortex studied by quantitative determinations of Purkinje cell RNA. II. RNA changes in rabbit cerebellar Purkinje cells after caloric stimulation and vestibular neurotomy. Acta Physiol. Scand., 67, suppl. 271, 1-24.

Kahan, B., Krigman, M.R., Wilson, J.E. and Glassman, E. 1970. Brain function and macromolecules. VI. Autoradiographic analysis of the effects of a brief training experience on the incorporation of uridine into mouse brain. Proc. Nat. Acad. Sci., 65, 300-303.

Luck, M.J., Rasmussen, P.S., Salaki, K. and Tsuetikov, K. 1958. Further studies on the fractionation of calf thymus histone. J. of Biol. Chem., 233, 1407-1414.

Machlus, B.J. 1971. Phosphorylation of nuclear proteins during behavior of rats. Ph.D. Dissertation, The Neurobiology Curriculum, The University of North Carolina, Chapel Hill, North Carolina, pp. 1-81.

Machlus, B.J., Wilson, J.E. and Glassman, E. 1971a. Effects of brief experiences on the phosphorylation of acid-extractable proteins in brain nuclei. Third International Meeting of the International Society for Neurochemistry, Budapest, Hungary.

Machlus, B.J., Wilson, J.E. and Glassman, E. 1971b. Phosphorylation of nuclear proteins during avoidance behavior of rats. The Society for Neuroscience First Annual Meeting, Washington, D.C.

Moore, S. and Stein, W.H. 1963. Chromatographic determination of amino acids by the use of automatic recording equipment. Methods in Enzymology, VI, 819-831.

Randerath, E. and Randerath, K. 1965. Ion exchange thin layer chromatography. XII. Quantitative elution and microdetermination of nucleoside monophosphates, ATP and other nucleotide coenzymes. Analyt. Biochem., 12, 83-93.

Rappoport, D.A. and Daginawala, H.F. 1968. Changes in nuclear RNA of brain induced by olfaction in catfish. J. Neurochem., 15, 991-1006.

Rosenzweig, M.R., Bennett, E.L. and Diamond, M.C. 1972. Chemical and anatomical plasticity of brain: Replications and extensions, 1970. In: Macromolecules and Behavior, J. Gaito (Ed.), Appleton-Century-Croft, New York, in press.

Schlesinger, K. and Wimer, R. 1967. Genotype and conditioned avoidance learning in the mouse. J. Comp. Physiol. Psychol., 63, 139-141.

Schwartz, J.H., Castellucci, V. and Kandel, E.R. J. Neurophysiol., in press.

Spackman, D.H., Stein, W.H. and Moore, S. 1958. Automatic record-
    ing apparatus for use in the chromatography of amino acids. Anal.
    Chem., 30, 1190-1206.
Stellwagen, R.H. and Cole, R.D. 1968. Danger of contamination in
    chromatographically prepared arginine-rich histone. J. of Biol.
    Chem., 243, 4452-4455.
Talwar, G.P., Chopra, S.P., Goel, B.K. and D'Monte, B. 1966. Corre-
    lation of the functional activity of the brain with metabolic
    parameters. III. Protein metabolism of the occipital cortex in
    relation to light stimulus. J. Neurochem., 13, 109-116.
Tsuboi, K.K. and Price, T.D. 1959. Isolation, detection and measure
    of microgram quantities of labeled tissue nucleotides. Arch.
    Bioch. Biophys., 81, 223-237.
Watson, W.E. 1965. An autoradiographic study of the incorporation
    of nucleic-acid precursors by neurones and glia during nerve
    stimulation. J. Physiol., 180, 754-765.
Weber, K. and Osborn, M. 1969. The reliability of molecular weight
    determinations by dodecyl sulfate-polyacrylamide gel electropho-
    resis. J. of Biol. Chem., 244, 4406-4412.
White, R.H. and Sundeen, C.D. 1967. The effect of light and light
    deprivation upon the ultrastructure of the larval mosquito eye.
    I. Polyribosomes and endoplasmic reticulum. J. Exp. Zool., 164,
    461-478.
Zemp, J.W., Wilson, J.E., Schlesinger, K., Boggan, W.O. and Glassman,
    E. 1966. Brain function and macromolecules. I. Incorporation
    of uridine into RNA of mouse brain during short-term training
    experience. Proc. Nat. Acad. Sci., 55, 1423-1431.
Zemp, J.W., Wilson, J.E. and Glassman, E. 1967. Brain function and
    macromolecules. II. Site of increased labeling of RNA in brains
    of mice during a short -term training experience. Proc. Nat. Acad.
    Sci., 58, 1120-1125.

FURTHER STUDIES ON MEMORY FORMATION IN THE GOLDFISH

Bernard W. Agranoff

Mental Health Research Institute, The University of

Michigan, Ann Arbor, Michigan

The concept of consolidation, or fixation, of memory is
historically linked to the effects of disruptive agents on the
brain and their interpretation. For example, retrograde amnesia
seen in man following physical trauma has been interpreted to indicate that permanent memory normally forms following a learning
experience. This idea was subsequently expanded by evidence for a
temporal gradient of growing insusceptibility to disruption of
newly-formed memory in both man and in experimental animals. Agents
producing convulsions and unconsciousness, such as electroconvulsive
shock, were reported to produce varying degrees of performance
decrement as a function of the time of their administration. The
question remained whether some unitary or multiple process of memory fixation existed or alternatively, whether the various disruptive agents had noxious properties that interfered with regis-
tration, storage or retrieval of behavioral information.

THE GOLDFISH

When we became interested in the biochemical aspects of brain
function, we began searching for a relatively simple animal model
to test effects of various agents on learning, and hopefully on
memory. After several frustrating months in which we attempted to
construct an automated apparatus that would measure learning in
various invertebrates, we turned to a relatively primitive vertebrate, the teleost fish. It has long been known that this cold-
blooded animal can be trained in the laboratory and remembers what
it has learned despite its relatively primitive brain. We found
that the common goldfish, Carrassius auratus, was readily available
in large numbers and for a fish, had a high ratio of brain to body

175

weight. A 10 gm goldfish has a brain weighing about 80 mg, enclosed
in a roomy cranial cavity such that it is possible to inject drugs
via a 30 gauge needle and a microsyringe in the unanesthetized
animal by penetrating the skull without danger of touching the
brain and with considerable assurance that the drug will rapidly
diffuse throughout the surrounding fluid and the brain (Agranoff
and Klinger, 1964). Thus, intracranial injection of 10 µl of a
solution containing 170 µg of puromycin rapidly produced inhibition
of fish brain protein synthesis by 80% (Agranoff, Davis and Brink,
1966). The inhibition was estimated by the change in incorporation
of injected labeled amino acids into protein. Recovery, with no
apparent ill effects, was within 18-24 hrs. Somewhat surprisingly,
during this period of inhibition, no gross neurological defects
such as circling, convulsions, etc., were observed.

TRAINING TASKS

We next examined the effect of injection of puromycin on
learning and memory in the goldfish by means of a shock-avoidance
task (Agranoff et al., 1966). An apparatus used for many of these
studies is shown in Figure 1. Goldfish are placed in individual
shuttleboxes. These consist of plastic tanks in which a barrier
has been placed such that the animal will ordinarily not swim over
it to the other side. The first trial begins with the onset of light
on the side of the apparatus in which the fish has been placed.
After 20 seconds, an intermittent punishing electrical shock is
administered through the water causing the fish to swim to the dark
side of the apparatus, presumably because it is safe-appearing.
This escape response requires no prior training. At the end of 40
sec, both light and shock turn off and the animal remains in the
dark side of the apparatus to which it escaped for the remaining
20 sec of the 1-min trial. At the beginning of the next min, the
light goes on on the second side of the apparatus and is followed
20 sec later by the intermittent shock as with the first trial.
The response of the animals is recorded by means of light beams
and photodetectors on either side of the barrier. The fish learn
to swim over the barrier before the onset of shock. They ordinarily
receive 20 trials on the first day of an experiment and are then returned to their home tanks. They are generally retrained 3-7 days
later. The details of the scoring of avoidance responses are stated
elsewhere (Davis, Bright and Agranoff, 1965; Agranoff, 1971). While
in Task I the animals are trained to swim to the dark side of the
apparatus upon a light signal, in Task II the mode is reversed so
that the onset of the trial is signaled by light on the opposite
side of the barrier, and the animal must swim to the light to
avoid the shock. More recently we have used Task III (Figure 2).
In this instance, the animal operates the shuttlebox. The
trial begins with the onset of light on the side of the apparatus
in which the animal has been placed, as in Task I. However, as the

TASK I & II

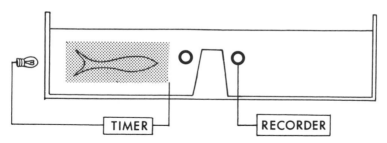

Figure 1.   Diagram of Task I for training of shock-avoidance.
Goldfish are trained to swim over the barrier on a light signal to
avoid a punishing electrical shock.   Light only for 20 seconds is
followed by light + shock for 20 seconds.   See text.   In Task II,
the light signal is on the opposite side of the box and the fish
is trained to swim into the light.

TASK III

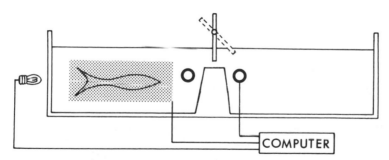

Figure 2.   Diagram of Task III.   Fish are trained to avoid shock,
but swimming over the barrier terminates the trial.   A clear plastic gate must be deflected in order to pass the barrier.

animal swims to the dark side of the apparatus, the light goes out,
terminating the trial.   The next trial begins one minute after the
first.   This task differs further from Tasks I and II in that in
addition to swimming over the barrier, the animal must deflect a
plastic gate with its snout, making the task somewhat more difficult
to learn.   The avoidance and escape latencies of each goldfish are
recorded by a PDP-8S computer in Task III, so that twenty shuttleboxes can be run simultaneously.   Our results with these three tasks
were similar except as noted below.

EFFECTS OF ANTIBIOTICS ON MEMORY

Using the antibiotic puromycin, we were led to the following conclusions (Agranoff, 1967):

1) Fish demonstrate acquisition (learning) of the shock avoidance task in a single 20-trial session and show good memory of what they have learned days to weeks later. In our experiments we deal with partial learning. At the end of 30 trials, fish are usually responding at about the 60% level. This has proved useful in studies on memory.

2) Puromycin injected just before or shortly following training results in memory loss, defined by a reduced responding rate on retesting several days later. The absence of such an effect of injections given several hours after training constitutes an important control experiment. It rules out the possibility that the poor performance seen on the retesting day could be due to lingering effects of the drug injection. The insusceptibility of memory to the antibiotic develops in about 1 hr with Tasks I and II, but only after 4 hrs in Task III.

3) When animals are injected prior to training, the rate of learning appears to be the same as that of animals not injected, or injected with saline. That aspect of memory associated with acquisition (short-term memory) is not blocked by the puromycin injection, while it appears that longer lasting memory is affected, since these animals, like those injected immediately after training, have a low avoidance-responding score.

4) When animals are trained and injected with puromycin immediately following training, we know that they will show no memory of learning when tested several days later. If these animals are, however, retrained shortly after training, they show normal avoidance-responding scores. The loss of memory does not occur instantaneously but develops slowly, over a period of 2-3 days. One can speculate that rather than blocking formation of permanent memory, the antibiotic injection accelerates forgetting. The posttrial injections appeared to reduce avoidance-responding rates to the naive level--that of untrained fish.

5) The question arose, Why did not at least some memory fixation occur during the initial training period? From an experiment that investigated possible effects of the environment on the onset of fixation, we learned that animals allowed to remain in the training apparatus for several hours following training remained susceptible to puromycin injections whereas had they been returned to their home tanks, they would have rapidly become insusceptible to the amnestic properties of the drug. Removal from a threatening

environment appeared to be necessary for the initiation of the
fixation process (Davis and Agranoff, 1966).

These findings prompted us to propose a multiphasic scheme
for the development of memory in which a short-term form is
converted to a long-term form of memory during a critical consolidation period which begins following training (Agranoff, 1967).
Such a model fits in well with the concept of retrograde amnesia,
but rather than requiring gross disruption of electrical activity
of the brain, it appears that more subtle tools are available. We
subsequently found that the glutaramide antibiotics, acetoxycycloheximide and cycloheximide produced effects similar to those described above with puromycin (Agranoff, Davis, Lim and Casola, 1968).
Actinomycin D, an RNA inhibitor, also produced these effects although
it was somewhat toxic (Agranoff, Davis, Casola and Lim, 1967).
Arabinosyl cytosine, an inhibitor of DNA synthesis had no effect
on either acquisition or on retention (Casola, Lim, Davis and
Agranoff, 1968). Of other substances tested, it is significant
that puromycin aminonucleoside (PAN) had no effect on learning or
memory. This substance resembles puromycin in structure and like
puromycin, lowers the threshold for pentylenetetrazol convulsions
(Agranoff, 1970). Its lack of effect on memory would seem to
dissociate the interactive effect of puromycin and PAN with a con-
vulsant, from the effect of puromycin on memory. The glutaramide
antibiotics and puromycin are quite different in structure and
appear to share chemically only an inhibitory effect on protein
synthesis, so that the most parsimonious interpretation of their
common action on memory formation appears to be on the basis of the
block in protein synthesis.

TEMPERATURE STUDIES

Because the goldfish is a poikilotherm, we have been able to
use this preparation to further explore the metabolic basis of a
consolidation process. We would expect a priori that the growing
insusceptibility to blocking agents with time following training
should, like other physiological processes, be temperature-depend-
ent. We had previously shown that susceptibility to electroconvulsive shock can be delayed by cooling fish (Davis, Bright and
Agranoff, 1965). In a study involving trout, Neale and Gray (1971)
were able to show that cooling trout for long periods of time did
not result in amnesia, even though amino acid incorporation into
protein is decreased. That is to say, a decrease in protein synthesis did not appear to be sufficient to block memory. Protein
synthesis must apparently be blocked while normal ongoing metabolism continues. We have recently examined further the effects
of temperature on consolidation in the goldfish using Task III
(Figure 3). In these experiments fish were given 30 trials on
Day 1 of an experiment and 10 additional trials 8 days later. Fish

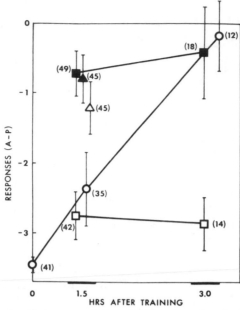

Figure 3. Effect of temperature following training on retention
measured on Day 8. Task III was used. In this experiment, fish
received 30 trials on Day 1 and 10 trials on Day 8. Retention
deficits were measured by means of a regression plot derived from
control groups run the same week as the corresponding experimental
animals.   O = Fish injected with 130 μg puromycin at times shown
and retrained on Day 8. Training and storage at 21–23°. □ , Fish
were trained, then stored at 14° for 1.5 or 3 hr, injected with
puromycin, then stored at 21–23° through retraining on Day 8. ■ ,
Fish cooled as above, but not injected. △ , Fish warmed to 30°C
for 1.5 hr after training, injected with 130 μg puromycin, then
stored at 21–23° through retraining on Day 8. ▲ , Fish warmed as
above, but not injected.

were trained at 21–30° and following training were either cooled
to 14° or heated to 30°. Control fish maintained at room temperature throughout the experiment showed a marked deficit when injected
with puromycin following training but not if they were injected 3
hr following training. Cooling animals to 14° immediately following training resulted in slowing of development of puromycin-insus-
ceptibility. In other words, cooling animals following training
delayed the onset of consolidation. More interestingly perhaps,
raising temperature of fish immediately following training resulted
in an acceleration in the growing insusceptibility of puromycin––
an apparent acceleration of consolidation. From other considerations, we know that both the cooling and the heating are aversive
to the fish. These results are all consistent with a metabolic
process underlying the consolidation of memory.

SIGNIFICANCE OF THE BEHAVIORAL EFFECTS OF THE ANTIBIOTICS

1. What is the Evidence for or against the Existence
of the Consolidation of Memory as a Multiphasic Process?

It has often been proposed that animals forget nothing. An
apparent loss of memory can be due to the development of an interfering process that blocks retrieval of information. At the behavioral level, it is possible that the antibiotics simply act as an
aversive stimulus, punishing the animal for what he has learned.
This is discussed below.

Several sorts of evidence from animal experiments have been
cited to show that animals who appeared to have lost memory actually
had a retrieval deficit. For example, it has been shown that rats
who had developed a memory loss after receiving various drugs or
electroconvulsive shock could remember what they had initially
learned if given an additional trial before testing (Quartermain,
McEwen and Azmitia, 1970). It has also been claimed that following
electroconvulsive shock, "lost" memory can reappear spontaneously
(Zinkin and Miller, 1967). In the case of studies with electroconvulsive shock and with the antibiotics in the fish, we have not
seen such recoveries. Supportive evidence for a multiphasic system
in formation of memory comes from an experiment in which blocking
or disrupting agents are not used. In this regard, Cherkin has
observed a biphasic memory both in chickens (1971) and in goldfish
(Riege and Cherkin, 1971). These results can be interpreted as
demonstrating the existence of a short-term memory with its decay,
a period of unavailability of forming memory, and its eventual
reappearance. Significantly, in order to demonstrate such a multiphasic retention curve, the stimulus must be mild.

2. Do the Antibiotics Produce Their Effects by Some
General Result of a Block in Macromolecular Synthesis
or via Some Few Selective Processes?

It is possible that agents that block RNA and protein synthesis in the brain can lead to a general lowering of brain effectiveness. For example, it might result in malaise or headache. It is
not readily apparent why such a putative global process would selectively block memory formation and have little effect on acquisition. To further examine this possibility, we trained fish and
then anesthetized them with the fish anesthetic tricaine methosul-
fate (MS-222, Finquel; Agranoff, 1971). Experimental groups were
in addition given puromycin intracranially 15 min after the anesthetic. Fish were allowed to recover after 4 hr and were retested
1 week later in Task III. Puromycin produced as big an effect, as
if the anesthetic had not been given, while the anesthetic alone
produced no effect. Puromycin given after recovery from the anesthetic produced no retention deficit on testing one week later.

The possibility that the inhibitors are exerting their effects at the level of neurotransmitter release or synthesis has been raised. For example it has been shown that puromycin blocks release of norepinephrine from peripheral nerves (Weiner and Rabadjija, 1968). Barondes and Cohen (1968) have reported that the cyclo-heximide-induced amnesia in the mouse can be partially blocked by the injection of amphetamines at critical times following train-ing. The result could equally well be interpreted as evidence for an action against some more global effect of both the inhibitors and the amphetamines on neuronal functioning as a whole.

3. If the Antibiotics Exert Their Effects via Macromolecular Synthesis, What is the Evidence for Changes in Macromolecular Synthe-sis with Learning or Memory Formation in the Absence of the Blocking Agents?

A simple model of brain function would provide for an increas-ing number of functional units such as synapses in the brain result-ing from behavioral experience. Such accretion hypotheses have not found much support. Although RNA is quite stable, RNA and protein turn over rapidly in the brain, and do not accumulate with age or with training. In an experiment in which animals were prelabeled via their pregnant mother with radioactive proteins, it could be shown that proteins were virtually all replaced within a few weeks of birth (Lajtha and Toth, 1966). By their nature, turnover studies in proteins require the use of isotopic methods and it is possible that small amounts of slowly turning-over proteins can exist unde-tected. In this regard, we have observed that axonally transported proteins in the goldfish tectum turn over more slowly than do brain proteins in general. Because of its completely crossed optic tracts, and the fact that the ganglion cells of the retina terminate in a specific region of the optic tectum, the goldfish has been the sub-ject of studies on transport in protein from the perikaryon to the presynaptic terminal as demonstrated by McEwen and Grafstein (1968). We have shown that several amino acids, particularly proline and asparagine are excellent precursors of axonally transported protein, since they generate very little systemic labeling (Elam and Agranoff, 1971). In a time study, we have shown that this labeled protein turns over with a half-life of several months. There is no evidence that visual stimulation increases this amount of protein nor is there neurophysiological or behavioral data to suggest that this sensory pathway is modified with behavioral change. However, the accessibility of both the cell body and the presynaptic terminals of the optic pathway may furnish us a model for less accessible parts of the nervous system such as the interneurons which may be better candidates for cells to be modified and mediate behavioral change.

That labeling of RNA may change as a result of learning experience in the goldfish was proposed by Shashoua (1968), who trained goldfish to swim upright following attachment of a styrofoam float to the lower jaw. He found that when labeled orotic acid was injected, radioactivity could subsequently be isolated from the resulting RNA in both uridylate and cytidylate moieties. The ratio of counts found in these two products was increased with training. We examined this phenomenon using our shuttlebox training (Baskin and Agranoff, 1971). We also found an increased U/C ratio albeit at a somewhat lower magnitude than had been reported. We established first that this increase in ratio was, in our hands, due to a decrease in C rather than to an increase in U. Since there is much more labeled U than labeled C whether animals are being trained or not, the decreased C did not contribute to a significantly decreased labeling of RNA. We subsequently learned that we could produce decreased labeled C in RNA in experiments in which animals were not acquiring the training task, i.e., "shock only" controls. We also were able to observe this phenomenon in fish placed in water previously occupied by stressed fish. Ultimately, we were able to show that the increase of the U/C ratio (decreased C) could be accounted for in our experiments by an increase in CO_2 in the water. Thus, the increased U/C ratio was not particularly related to training, nor was it related to synthesis of a novel RNA, since we were able to establish further that the decrease in RNA labeling was due to a change in the cytidylate precursor pool. We have not as yet established the mechanism for this lowering, but conversion of UTP to CTP requires ATP and glutamine, and it is possible that CO_2 affects cerebral blood flow in a way that diminishes either or both of these substrates. We have thus not observed alterations in RNA or protein synthesis in the fish that would lead us to an independent confirmation of macromolecular involvement in learning or memory.

PROSPECTS

Further directions in chemical approaches to memory appear to diverge. On the one hand it is necessary to seek simpler models. Shock avoidance involves sensory, motor and motivational components. It is of great interest to the biologist to see if he can examine the various components of behavior and establish whether they are also blocked by the antimetabolites. For example, it is important to know whether emotionality of an animal is altered as a result of antimetabolite injections. Experiments of a previous generation attempting to localize regions of brain involved in learning were notably unsuccessful. Yet, electrophysiological, morphological and behavioral techniques are more sophisticated today and further approaches in this direction may not be so fruitless. On the other hand, from the standpoint of applicability, we would like to know if what we see in the fish and the rat is also true in the monkey,

and ultimately in man. Selective effects on memory often seen in
the aged underscore the need for better understanding at a molecular level of the normal and pathological events that mediate memory
formation and storage.

ACKNOWLEDGMENT

    The author gratefully acknowledges the skillful assistance of
Mr. Paul Klinger in the reported experiments. This research was
supported by grants from the National Institute of Mental Health
and the National Science Foundation.

REFERENCES

Agranoff, B.W. 1967. Agents that block memory. In: The Neurosci-
    ences: A Study Program, G.C. Quarton, T. Melnechuk and F.O. Schmitt
    (Eds.), The Rockefeller University Press, New York, pp 756-764.
Agranoff, B.W. 1970. Protein synthesis and memory formation. In:
    Protein Metabolism of the Nervous System, A. Lajtha (Ed.), Plenum
    Press, New York, pp. 533-543.
Agranoff, B.W. 1971. Effects of antibiotics on long-term memory
    formation in the goldfish. In: Animal Memory, W. K. Honig and
    P.H.R. James (Eds.), Academic Press, New York, pp. 243-258.
Agranoff, B.W., Davis, R.E. and Brink, J.J. 1966. Chemical stud-
    ies on memory fixation in goldfish. Brain Res., 1, 303-309.
Agranoff, B.W. and Klinger, P.D. 1964. Puromycin effect on memory
    fixation in the goldfish. Science, 146, 952-953.
Agranoff, B.W., Davis, R.E., Casola, L. and Lim. R. 1967. Actino-
    mycin D blocks memory formation of a shock-avoidance in the gold-
    fish. Science, 158, 1600-1601.
Agranoff, B.W., Davis, R.E., Lim, R. and Casola, L. 1968. Biolo-
    gical effects of antimetabolites used in behavioral studies. In:
    Psychopharmacology: A Review of Progress, 1957-1967, D.E. Efron,
    J.O. Cole, J. Levine and J.R. Wittenborn (Eds.), Public Health
    Service Publication No. 1836, pp. 909-917.
Barondes, S.H. and Cohen, H.D. 1968. Arousal and the conversion
    of "short-term" to "long-term" memory. Proc. Nat. Acad. Sci.,
    61, 923-929.
Baskin, F. and Agranoff, B.W. 1971. Effect of various stresses on
    incorporation of ^{3}H-orotic acid into goldfish brain RNA. Trans.
    Am. Soc. Neurochem., 2, 56.
Casola, L., Lim, R., Davis, R.E. and Agranoff, B.W. 1968. Beha-
    vioral and biochemical effects of intracranial injection of cyto-
    sine arabinoside in goldfish. Proc. Nat. Acad. Sci., 60, 1389-
    1395.
Cherkin, A. 1971. Biphasic time course of performance after one-
    trial avoidance training in the chicken. Commun. Behav. Biol.,
    5, 379-381.
Davis, R.E. and Agranoff, B.W. 1966. Stages of memory formation

in goldfish: Evidence for an environmental trigger. <u>Proc</u>. <u>Nat</u>. <u>Acad</u>. <u>Sci</u>., <u>55</u>, 555-559.

Davis, R.E., Bright, P.J. and Agranoff, B.W. 1965. Effect of ECS and puromycin on memory in fish. <u>J</u>. <u>Comp</u>. <u>Physiol</u>. <u>Psychol</u>., <u>60</u>, 162-166.

Elam, J.S. and Agranoff, B.W. 1971. Rapid transport of protein in the goldfish optic tectum. <u>J</u>. <u>Neurochem</u>., <u>18</u>, 375-387.

Lajtha, A. and Toth, J. 1966. Instability of cerebral proteins. <u>Biochem</u>. <u>Biophys</u>. <u>Res</u>. <u>Comm</u>. <u>23</u>, 294-298.

McEwen, B.S. and Grafstein, B. 1968. Fast and slow components in axonal transport of protein. <u>J</u>. <u>Cell</u> <u>Biology</u>, <u>38</u>, 494-508.

Neale, J.H. and Gray, I. 1971. Protein synthesis and retention of a conditioned response in rainbow trout as affected by temperature reduction. <u>Brain</u> <u>Res</u>., <u>26</u>, 159-168.

Quartermain, D., McEwen, B.S. and Azmitia, E.C., Jr. 1970. Amnesia produced by electroconvulsive shock or cycloheximide: Conditions for recovery. <u>Science</u>, <u>169</u>, 683-686.

Riege, W.H. and Cherkin, A. 1971. One-trial learning and biphasic time course of performance in the goldfish. <u>Science</u>, <u>172</u>, 966-968.

Shashoua, V.E. 1968. RNA changes in goldfish brain during learning. <u>Nature</u>, <u>217</u>, 238-240.

Weiner, N. and Rabadjija, M. 1968. The regulation of norepinephrine synthesis. Effect of puromycin on the accelerated synthesis of norepinephrin dissociated with nerve stimulation. <u>J</u>. <u>Pharm</u>. <u>Exp</u>. <u>Ther</u>., <u>164</u>, 103-114.

Zinkin, S. and Miller, A.J. 1967. Recovery of memory after amnesia induced by electroconvulsive shock. <u>Science</u>, <u>155</u>, 102-103.

THE CHOLINERGIC SYNAPSE AND THE SITE OF MEMORY

J. Anthony Deutsch

Department of Psychology, University of California at
San Diego, La Jolla, California

The idea that learning and memory are due to some form of change of synaptic conductance is very old, having been suggested by Tanzi in 1893. It is a simple idea and in many ways an obvious one. However, the evidence that learning is due to changes at the synapse has hitherto been meager (Eccles, 1961, 1964; Spencer and Wigdor, 1965; Beswick and Conroy, 1965; Fentress and Doty, 1966). Though changes do occur at a spinal synapse as a result of stimulation, there is no evidence that the changes are those utilized in the nervous system for information storage. To use an analogy, if we pass large amounts of current across resistors in a computer, temporary increases in temperature and perhaps even permanent increases in resistance occur. However, such an experiment shows only that the computer could store information by using "post-stimulation" alterations in its resistors but not that this is the actual way in which the computer does store information. Further, Sharpless (1964) has pointed out that learning is not due to simple use of stimulation of a pathway and he therefore questions whether the phenomena studied by Eccles (1961, 1964) have anything to do with learning as observed in the intact organism. Nevertheless this does not mean that learning is not due to synaptic changes of some sort. It means only that a different experimental test of the possibility must be devised.

In designing our experimental approach to this problem clues from human clinical evidence were used. After blows to the head sustained in accidents, events which occurred closest in time prior to the accident cannot be recalled (retrograde amnesia). Such patches of amnesia may cover days or even weeks. The lost memories tend to return with those most distant in time from the accident becoming

187

available first (Russell and Nathan, 1946). In the Korsakoff
syndrome (Talland, 1965), retrograde amnesia may gradually increase
until it covers a span of many years. An elderly patient may end
up remembering only his youth, while there is no useful memory of
the more recent intervening years. From such evidence concerning
human retrograde amnesia we may conclude that the changes that
occur in the substrate of memory take a relatively long time and
are measurable in hours, days and even months. If we suppose from
this that the substrate of memory is synaptic and that it is slowly
changing, then it may be possible to follow such synaptic changes
using pharmacological methods. If the same dose of a synaptically
acting drug has different effects on remembering depending on the
age of the memory (and this can be shown for a number of synaptically acting drugs) we may assume that there has been a synaptic
alteration as a function of time since learning and we may infer
that such a synaptic change underlies memory.

Pharmacological agents are available which can either increase
or decrease the effectiveness of neural transmitters (Goodman and
Gilman, 1965). For instance, anticholinesterase and anticholinergic
drugs affect transmission at synapses utilizing the transmitter
acetylcholine. During normal transmission, acetylcholine is rapidly
destroyed by the enzyme cholinesterase. Anticholinesterase drugs
such as physostigmine and diisopropyl fluorophosphate (DFP) inactivate cholinesterase, and so indirectly prevent the destruction of
acetylcholine. In submaximal dosage these drugs inactivate not all
but only a part of the cholinesterase present and hence only slow
down but not stop the destruction of acetylcholine. The overall
effect at such submaximal levels of anticholinesterase is to increase by some constant the lifetime of any acetylcholine emitted
into the synapse and to increase thereby the acetylcholine synaptic
concentrations resulting from a given rate of emission. Up to a
certain level the greater this concentration the greater is the efficiency of transmission, i.e., the conduction across the synapse.
Above that level, which is set by the sensitivity of the postsynaptic membrane, any further increase in acetylcholine concentration
produces a synaptic block (Goodman and Gilman, 1965; Feldberg and
Vartiainen, 1934; Volle and Koelle, 1961). Thus the application
of a given dosage of anticholinesterase will (by protecting acetylcholine from destruction) have different effects on the efficiency
of synaptic conduction depending on the rate of acetylcholine emission during transmission and on the sensitivity of the postsynaptic
membrane. At low levels of emission of acetylcholine or low sensitivity of the postsynaptic membrane an application of anticholinesterase will render transmission more efficient. Such a property is
used to good effect in the treatment of myasthenia gravis. In the
treatment of this disorder anticholinesterase is used to raise the
effective concentration of acetylcholine at the neuromuscular junction and so to reduce apparent muscular weakness. On the other
hand the same dose of acetylcholinesterase that caused muscular

contraction in the myasthenic patient produces paralysis in a man with normal levels of function at the neuromuscular junction.

If there are changes with time after learning in the level of acetylcholine emitted at the modified synapse, then such a synapse should show either facilitation or block depending on when in time after learning we inject the same dose of anticholinesterase. A similar argument with regard to the action of anticholinesterase can be applied if we assume that instead of a presynaptic increment in transmitter, it is the postsynaptic membrane which becomes more sensitive to transmitter as a function of time after learning. But the use of an anticholinesterase does not allow us to decide which of these alternative versions of the hypothesis of the increment of synaptic conductance actually holds for the learning situation. Later, however, I shall indicate how the use of other types of drugs such as the cholinomimetics allows us to surmise that postsynaptic sensitization is the more likely mechanism.

The first two experiments (Deutsch, Hamburg and Dahl, 1966; Deutsch and Leibowitz, 1966) show that facilitation or block of a memory can be obtained with the same dose of anticholinesterase simply as a function of time of injection since original learning as might be expected if synaptic change formed the substrate of memory.

EXPERIMENTAL INVESTIGATIONS

In the first experiment, rats were trained on a simple task (Sprague Dawley males approximately 350 g at the start of the experiment). Then an intracerebral injection of anticholinesterase was made at different times after initial training, the time being varied from one group of subjects to another. After injection, all rats, irrespective of the group to which they were assigned, were retested 24 hours after injection. Thus, what was varied was the time between training and injection. The time between injection and retest was kept constant. Any difference in remembering between groups was therefore due to the time between initial training and injection.

Rats were placed on an electrified grid in a Y-maze. The lit arm of the Y was not electrified and its position was changed randomly from trial to trial. The rats therefore learned to run into the lit arm. The criterion of learning was met when they had chosen the lit arm 10 trials in succession whereupon training was concluded.

Then at various times after training, the rats were injected intracerebrally with DFP dissolved in peanut oil. The subjects were placed in a stereotaxic instrument under nembutal anesthesia.

They were intracerebrally injected in two symetrically placed bilateral loci. The placements were: anterior 3, lateral 3, vertical
+2 and anterior 3, lateral 4.75, vertical -2, according to the
atlas of DeGroot (1957). 0.01 ml of peanut oil containing 0.1 per
cent of diisopropyl fluorophosphate (DFP) was injected in each
locus. This dose did not increase the number of trials to criterion
in a naive group of rats thus showing that learning capacity during
training was not affected by the drug in the amounts used. At 24
hours after injection, the rats were retrained to the same criterion of 10 successive trials correct. The number of trials to criterion in this retraining session represented the measure of retention.

        The first group was injected 30 minutes after training. Its
retention was significantly worse than that of a control group
injected only with peanut oil (Except as otherwise stated, the
results quoted are significant beyond the 1% level. The tests used
were the t-test, Mann-Whitney U test and analysis of variance). By
contrast, a group injected with DFP 3 days after training showed
the same amount of retention as did the control group. Thus, up
to this point it seems that memory is less susceptible to DFP the
older it is. In fact, a subsidiary experiment (Deutsch and Stone,
unpublished) has established that injections of DFP on habits 1 and
2 days old have no effect, showing that the initial stage of vulnerability lasts less than 1 day. Beyond 3 days, however, the
situation seems to reverse itself: the memory is more susceptible
to DFP the older it is because a DFP group injected 5 days after
training, showed only slight recollection at retest, and a further
group injected 14 days after training, showed complete amnesia. The
score of the group trained 14 days before injection was the same as
the score of the previously mentioned naive group which has not
been trained before but had simply been injected with DFP 24 hours
prior to testing. The amnesia of the DFP group trained 14 days
before injection was not due to normal forgetting, since other controls showed almost perfect retention over a 15 day span. Similar
results have been obtained by Hamburg (1967) with intraperitoneal
injections of the anticholinesterase physostigmine, using the same
escape habit. Biederman (1970) confirmed the shape of the amnesic
function with physostigmine in an operant situation. He used a
latency measure of forgetting and a bar press response.

        To make sure that we were not observing some periodicity in
fear or emotionality interacting with the drug, another experiment
employing an appetitive rather than an escape task was conducted.
The rats were taught to run a reward of sugar-water the position of
which always coincided with the lit arm of a Y-maze (Wiener and
Deutsch, 1968). As seen in Figure 1, the results when compared to
the maze results from the preceding experiments show a very similar
pattern of amnesia as a function of time of learning before injection.

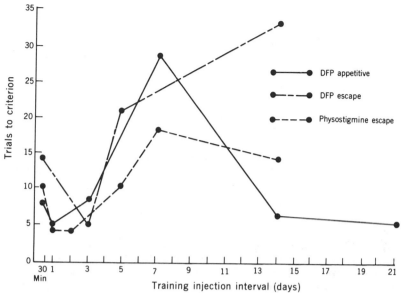

Figure 1. The effect of anticholinesterase injection on memories of different age, shown in three separate experiments. Trials to criterion during retest are plotted against the time which elapsed between retest and original learning. A larger number of trials to criterion during retest signifies a greater amnesia. The time between injection and retest was constant. The differences past the 7 day point probably represent differing rates of forgetting in the three situations. The three experiments are Deutsch et al. (1966), Hamburg (1967), and Wiener and Deutsch (1968).

It is therefore most likely that we are in fact studying memory. The divergences in the curves after 7 days are probably due to differences in rates of forgetting among the three groups.

In this first set of experiments which dealt with the effects of the anticholinesterases DFP and physostigmine on habits which are normally well retained, the effects of these drugs were to decrease the retention of a habit depending on its age. Thus, one of the predicted effects of an anticholinesterase was verified. However, the other predicted effect, facilitation, was not shown. The reason for this is that the habit which was acquired was so well retained without treatment over 14 days that one could not, on methodological grounds, show any improvement of retention subsequent to injection of the drug. It may be the case that 1, 2 and 3 day old habits were facilitated instead of merely being unaffected, but the design of the experiment would not allow us to detect this because there is an effective ceiling on performance. Therefore, an attempt was made to obtain facilitation where it was

methodologically possible to detect it, namely, where retention of
the habit by a control group was imperfect. For example, it was
found that 29 days after learning the escape habit described above
was almost forgotten by a group of animals injected with peanut oil
only 24 hours before. On the basis of this observation, a second
kind of experiment was devised.

Rats were divided into four groups. The first two groups were
trained 14 days before injection, the second two groups 28 days
before injection. One of the 28-day and the 14-day groups were
injected with the same dose of DFP, the remaining 28-day and 14-day
groups were injected with the same volume of pure peanut oil instead. The experimental procedure and dosage were exactly the same
as previously described.

On retest, poor retention was exhibited by the 14-day DFP group
and 28-day peanut oil group. By contrast, the 28-day DFP group and
the 14-day peanut oil group exhibited good retention. The results
of anticholinesterase injection show a large and clear facilitation
of an otherwise almost forgotten 28-day old habit while they confirm
the obliteration of an otherwise well-remembered 14-day old habit
already demonstrated in the previous experiments (Figure 2a). The
same facilitation of a forgotten habit was shown by Wiener and
Deutsch (1968) using an appetitive habit and by Squire (1970) using
physostigmine-injected mice. Biederman (in press) showed an improvement in memory in pigeons when physostigmine is injected 28
days after a line tilt discrimination was partly learned. A well-
learned color discrimination acquired by the same subjects showed
no such improvement under the same conditions. Thus, these results
also lend strong support to the notion that forgetting is due to a
reversal of the change in synaptic conductance which underlies
learning (Figure 2b). It must be emphasized, however, that both
the block and facilitation of a memory are temporary, wearing off
as the injected drug wears off.

So far it has been shown that the anticholinesterase drugs DFP
and physostigmine have different effects on memories of different
age. Though their actions on memory are consistent with, and plausibly interpreted by their anticholinesterase action, some other
property besides their indirect action on acetylcholine could in
some unknown manner produce the same results. It was, thus, desir-
able to conduct an independent check on the hypothesis that the
effects observed are due to an effect on acetylcholine. This check
can be provided by the use of an anticholinergic drug. An anticholinergic drug (like atropine or scopolamine) reduces the effective
action of a given level of acetylcholine at the synapse without
actually changing the level itself. It does this apparently by
occupying some of the receptor sites on the postsynaptic membrane
without producing depolarization. It thus prevents acetylcholine

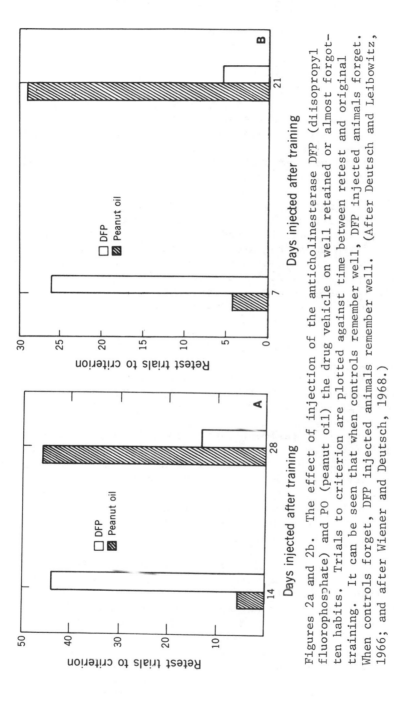

Figures 2a and 2b. The effect of injection of the anticholinesterase DFP (diisopropyl fluorophosphate) and PO (peanut oil) the drug vehicle on well retained or almost forgotten habits. Trials to criterion are plotted against time between retest and original training. It can be seen that when controls remember well, DFP injected animals forget. When controls forget, DFP injected animals remember well. (After Deutsch and Leibowitz, 1966; and after Wiener and Deutsch, 1968.)

from reaching such receptor sites and so attenuates the effectiveness of this transmitter. We would therefore expect an anticholinergic to block conduction at a synapse where the postsynaptic
membrane is relatively insensitive, while simply diminishing conduction at synapses where the postsynaptic membrane is highly sensitive. If the interpretation of the effects of DFP is correct,
we would then expect the reverse effect with the administration of
an anticholinergic drug. That is, we would expect the greatest
amnesia with anticholinergics precisely where the effect of anticholinesterase was the least; and we would predict the least effect
where the effect of anticholinesterase on memory was the largest.
It will be recalled that the least effect of anticholinesterase was
on habits 1 to 3 days of age.

In a third set of experiments (Wiener and Deutsch, 1968;
Deutsch and Rocklin, 1967) the anticholinergic agent employed was
scopolamine, and it was injected using exactly the same amount of
oil and location as in the previous experiments using DFP. Deutsch
and Rocklin used an injection of scopolamine at the same loci as in
the first experiment. Peanut oil (0.01 ml) containing 0.58 percent
of scopolamine was injected in each placement. Wiener and Deutsch
used only the first locus, but doubled the amount injected at that
site (both of scopolamine and DFP). The same experimental procedure was also used. A group injected 30 minutes after training
showed little if any effect of scopolamine. However, a group
injected 1 and 3 days after training showed a considerable degree
of block. Groups injected 7 and 14 days after training showed
little if any effect. The results from the appetitive and escape
situations were very similar.

As far as the experimental methodology will allow us to dis-
cern, the effect, then, of an anticholinergic is the mirror-image
of the anticholinesterase effect (Figure 3). There is an increase
of sensitivity between 30 minutes and 1-to-3 days, followed by
a decrease of sensitivity. This further confirms the notion that
there are two phases present in memory storage. Finally, it is of
interest to note that amnesia can result in man from anticholinergic therapy (Cutting, 1964).

The experiments already outlined support the idea that at
the time of learning some unknown event stimulates a particular
group of synapses to alter their state and to increase their conductivity. At this point we may ask why such an increase in synaptic conductivity does not manifest itself with the passage of time
when no drugs are injected. Why has it not been noted that habits
are better remembered a week after initial learning than, say,
three days after such learning? There are various possible answers.
One is that the phenomena we have described are some artifact of
drug injection. Another is that animal training has, in general,

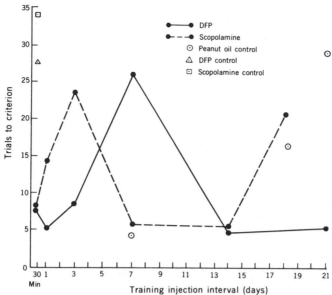

Figure 3. The effects of the injection of the anticholinergic
scopolamine compared with that of the anticholinesterase DFP and
control injections of PO (peanut oil) on the retention of an appetitive task at various times after original learning. The time
between injection and retest was constant. Also indicated are the
number of trials to criterion when rats were injected with scopolamine (CTL scopolamine) of DFP (CTL DFP) before original learning
to give an estimate of actual amount of amnesia produced. (From
Wiener and Deutsch, 1968.)

stretched over days in other studies, blurring in time the initiation of a memory. In addition, and partly as a consequence of the
foregoing, it is difficult to find studies where the age of the
habit, measured in days, has been used as an independent variable
in studies of retention. However should we not have seen such an
improvement in recall in our control groups? This would have been
unlikely for the methodological reasons that our animals were
trained to the very high criterion of 10 out of 10 trials correct.
Given a score which was initially almost perfect, it was thus well
nigh impossible to observe any subsequent improvement in retention
that might in fact actually exist. To rid ourselves of this methodological limitation, we devised a study in which rats were initially undertrained using escape from shock. The rats were given
15 trials. We then waited to see how many trials it would take
these rats on some subsequent day to reach our strict criterion
(Huppert and Deutsch, 1969). No drugs were used. We found that
the rats took only about half the number of trials to reach criterion
when they waited 7 or 10 days than when they waited 1 or 3 days

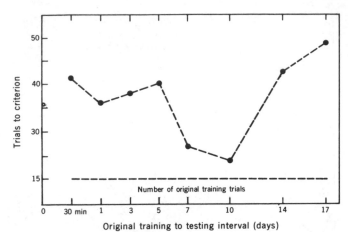

Figure 4. The effects of delay between original partial training
(15 trials) and subsequent training to criterion. Plotted are
trials to criterion in subsequent training against time since
original partial training. Control indicates the number of trials
to criterion taken by a group which received its training all in
one session.

(Figure 4). Huppert (personal communication) has now shown an
analogous improvement using an appetitive task. Finally, Dr. J. L.
McGaugh has pointed out to me that there are old animal studies
which purport to find similar effects (Anderson, 1940; Bunch and
Magdsick, 1933; Bunch and Lang, 1939; Hubbert, 1915). This shows
that our conclusions about the varying substrate of memory were
not due to some pharmacological artifact.

   We may now ask ourselves whether the inferred modification of
a synapse represents an all-or-none or a graded process. In other
words can a synapse be modified only once during learning or does
a repetition of the same learning task after some learning has
already occurred further increase conductance at a single synapse.
If we postulate an all-or-none process then how according to such
a model can we explain empirical increases in "habit strength" with
increased training? Possibly they are due to a progressive involvement of fresh synapses and a spread involving more parallel con-
nections in the nervous system. In support of a graded process,
we may hypothesize that successive learning trials modify the same
synapses in a cumulative way by producing an increase either in the
rate at which conductance increases or in the upper limit of such
conductance or both.

   There are tests of these two alternatives. If, with increased
training a synapse becomes more conductive, then a habit should be-

come increasingly more vulnerable to anticholinesterase with increased training. Furthermore, the memory of the same habit should be facilitated when its level of training is very low. In other words, we should be able to perform the same manipulations of memory by varying level of training as we were already able to perform when we varied time since training.

If, on the other hand, increases in training simply involve a larger number of synapses but no increase in the level of transmitter at any one synapse, then increases in training should not lead to an increased vulnerability of a habit to anticholinesterase. Rather, the opposite should be the case. As the number of synapses recruited is increased, some of the additional synapses will, by chance variation, be less sensitive to a given level of anticholinesterase. Thus, a larger number of synapses should be left functional after anticholinesterase injection when we test an overtrained habit. Three experiments (Deutsch and Leibowitz, 1966; Deutsch and Lutzky, 1967; Leibowitz, Deutsch and Coons, in preparation) show a large and unequivocal effect. Poorly learned habits are enormously facilitated and well learned habits are blocked (Figure 5). This supports the hypothesis that a set of synapses underlying a single habit remains restricted and each synapse within such a set simply increases in conductance as learning proceeds.

So far the results presented have been interpreted in terms of the action of drugs on synapses which alter their conductance as a function of time since training and amount of training. We can use the model we have developed to generate a somewhat different kind of prediction. An anticholinesterase in submaximal concentrations simply slows down the rate of destruction of acetylcholine. Since we have hypothesized that amnesia is due to a block resulting from an acetylcholine excess, we should predict no amnesia if we spaced our trials so that all or most of the acetylcholine emitted on the previous trial is destroyed by the time the next trial comes along. It has been shown by Bacq and Brown (1937) that (with an intermediate dose of anticholinesterase) block at a synapse occurred only when the intervals between successive stimuli were shortened. Accordingly, an experiment was performed where we varied the interval during retest between 25 sec and 50 sec (Rocklin and Deutsch, unpublished). Using a counterbalanced design it was found that rats tested under physostigmine at 25 sec intervals showed amnesia for the original habit. Those tested at a 50 sec intertrial interval under physostigmine showed no amnesia.

In a second experiment the rats during retest had to learn an escape habit reverse of the one they had learned during training. Therefore, to escape shock during retest they had to learn not only to run to the dark alley but also to inhibit the original learning

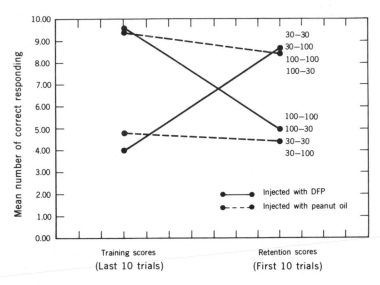

Figure 5.   The effects of anticholinesterase injection (DFP) on the
retention of well-learned and poorly learned habits.   The mean
number of correct responses of the last 10 of 30 trials for two
groups are shown on the left.   One group had to learn to run to
alley illuminated by bulb with 30 V; the other had to learn the
same task except that the bulb had 100 V across it.   As can be
seen from the last 10 trials, the dim light of the 30 V group posed
a difficult task which produced little learning by the end of the
30 trials.   The group learning the brighter cue (100 V) displayed
excellent acquisition.   Because of the different rates of acquisition of the 100 V and 30 V habits, half of each group was shifted
to retest on the other brightness and half was retrained on the
same brightness (30-30, 100-100 retested on the same brightness,
30-100 trained on 30, retested on 100, 100-30 trained on 100,
retested on 30).   The scores of animals trained on the same brightness are combined.   Half the animals were injected with DFP, the
other half with peanut oil (PO).   There is little change in the
scores of the peanut oil animals.   However, there is a complete
crossover of the drug injected animals, showing block of the well-
learned habit and facilitation of the poorly learned habit.

of running to the lit alley.   Thus, provided that the original habit
was remembered at the time the reversal was being learned, the time
to learn the reversal should take longer than the time to learn the
original habit.   But if the original habit was not remembered there
should be no difference in trials to criterion between original
learning and retest.   The results showed that at 50 sec between
trials animals in both the physostigmine and the saline control
groups took almost twice as long to reverse as it took them to learn

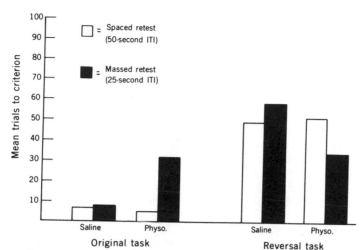

Figure 6. The effect of massing and spacing trials during retest
on anticholinesterase-induced amnesia. On the left, retest consisted of relearning original habit (run to light, avoid dark).
On the right, retest consisted of unlearning original habit. On
retest the animal had to learn to run to dark and avoid light
(reversal). (ITI = intertrial interval, PHYSO = physostigmine).

TABLE I

Treatment	Median Number of Trials to Criterion	
	3 days	7 days
Carbachol	6.0 (15)	20 (15)*
Saline	4.0 (8)	0 (7)

*$P < .01$ compared with saline, Mann-Whitney U test.

The effect of carbachol injection on recall of habits that were 3
and 7 days old. Criterion was seven correct trials in succession.
Numbers in parentheses indicate number of rats tested.

the original habit, indicating in fact that they remembered the original habit (Figure 6). At 25 sec. between trials, the physostigmine animals learned the reversal as quickly as the original habit whereas again the saline animals took much longer. This second experiment shows that the amnesia of the 25 sec physostigmine group in the first experiment is not due to disorientation or an incapacity to perform or learn but to an amnesia. We might explain the high relearning scores of the same habit of the rats run at 25 sec. intervals under physostigmine if they had to run at 25 sec intervals. However, it is difficult to see how such incapacitation could produce abnormally low learning scores of the reversal habit. This dependence of the amnesia on the precise interval of trials during retest should of course not be seen with anticholinergics or cholinomimetics but only with ancholinesterases. This further prediction from the hypothesis should be tested.

So far, then, it seems that the drugs we are using to block or facilitate memory have their effect on synaptic conductance. However, what is it that changes when synaptic conductance alters? As mentioned previously, the two main hypotheses are (a) that the amount of transmitter emitted at the presynaptic ending increases or (b) that the postsynaptic ending increases in its sensitivity to transmitter. To test this idea, carbachol (carbamoylcholine) was injected before retest. This drug is a cholinomimetic. It acts on the postsynaptic membrane much like acetylcholine. However, it is not susceptible to destruction by the enzyme acetylcholinesterase. Therefore, by injecting this drug, we can test the sensitivity of the postsynaptic membrane. It seems that 7 day old habits are blocked by a dose of this cholinomimetic which leaves a 3-day old habit unaffected (Table I). This would indicate that it is probably the postsynaptic membrane that has increased its sensitivity and so increased synaptic conductance.

One of the questions that often arises is why it is that we do not block all cholinergic synaptic activity with the drugs we use. As was seen above rats learn appetitive tasks at a normal rate under doses of drug which under some circumstances produce complete amnesia. There is very little in the overt behavior of the rat to indicate that it has been drugged. The doses of drugs used produce no apparent malaise or uncoordination. Clearly, the dose we use only seems to affect what one might call the "memory" synapses. It would seem that these are, therefore, more sensitive to our drugs. Such an abnormal sensitivity may be more apparent than real. We know that there are some levels of training and times after training where a habit is unaffected by the dosage of drug we use, and this shows that "memory" synapses are not always affected. It therefore seems more plausible to think of the "memory" synapses as traveling through a much larger range of postsynaptic sensitivity, while normal synapses remain fixed somewhere in the middle of the range of sensitivity variation of the memory synapse. In other words, the "memory" synapse has to swing from extreme insensitivity to transmitter to extreme sensitivity in order to manifest those changes in conductance which we have demonstrated. It will therefore be much more susceptible to anticholinergic agents when conductance is low and to anticholinesterases and cholinomimetics when conductance is high. In the middle of the range, sensitivity to all agents will resemble that of a normal synapse and only grossly toxic doses will affect memory. This speculation, of course, will have to be further tested. The experiments so far reported implicate the cholinergic system in memory. It is, of course, possible that other systems such as the adrenergic will also turn out to have a similar function, and this, too, we hope to test.

When an animal is rewarded for performing a habit such a habit will be learned or acquired. However, when the habit is no longer rewarded, the animal will cease to perform the habit. Another kind of learning takes place, and this is called extinction. If initial learning consists of the formation of some synaptic (or other) connection, does extinction consist of the weakening or uncoupling of this connection? Or is it the formation of some other connection which then works to oppose the effects of the first ("learning") connection? If extinction consists of weakening the connection set up in original learning, then an extinguished habit should be similar to a forgotten habit pharmacologically. We have already shown that an almost forgotten habit is facilitated by anticholinesterase. We would, then, on the "weakening" hypothesis of extinction, expect an injection of an anticholinesterase to produce less amnesia of an extinguished habit than of the same unextinguished habit. If, on the other hand, during extinction there is another habit acquired which inhibits the expression of the original habit, another pattern of results should be discernible after injection with an anticholinesterase. If original learning occurs 7 days before anticholinesterase injection and retest, there should be amnesia for the original habit. If extinction of the habit is given close in time to its acquisition, there should be amnesia for both the original learning and extinction. If, on the other hand, original learning is 7 days before injection and retest, the extinction is 3 days before injection and retest, the original habit should be lost but the extinction habit retained (As we noted above, 3 day old habits are unaffected by our dose of anticholinesterase). When extinction was given to rats close in time to the original training, both the original training and extinction were blocked by physostigmine (Deutsch and Wiener, 1969). Those rats took the same number of trials to relearn as control animals, which were trained, not extinguished and then drugged. However, when extinction was placed 3 days before injection and retest it took the rats during retest after drug injection approximately twice as many trials to learn as control animals (unextinguished and drugged), showing that

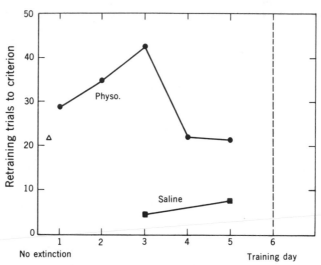

Figure 7. The effect of physostigmine on retraining after extinction. The time between original learning and retraining is the
same for all groups. Where time of extinction is close to original
learning, there is amnesia but no difference from group receiving
no extinction. At extinction 3 days before learning, the number of
trials to relearn is almost double (SALINE - scores of controls
injected with saline, PHYSO - scores of animals injected with
physostigmine).

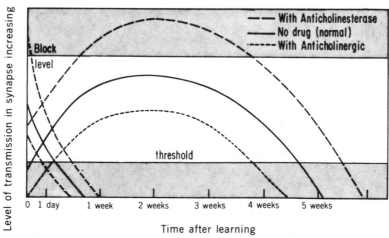

Figure 8. The hypothesized changes in "memory" synapses with time
after training and with pharmacological intervention.

extinction is the learning of a separate habit opposing the performance of the initially rewarded habit.

It has also been suggested (Carlton, 1969) that different systems such as excitatory or inhibitory systems are subserved by
different transmitters. Habits acquired during extinction have
been viewed as inhibitory. However, the last experiment we have
outlined also shows that extinction placed close to original learning is equally as vulnerable to anticholinesterase as original
learning. Habits can probably not be classified into synaptically
inhibitory and excitatory on the basis of behavioral excitation or
inhibition. However, as all habits compete for behavioral expression, there must be excitation and reciprocal inhibition connected
with all habits.

CONCLUSIONS

A simple hypothesis can explain the results obtained to date
if we disregard those results when we wait 30 minutes after original
learning to inject. The hypothesis is that, as a result of learning,
the postsynaptic endings at a specific set of synapses become more
sensitive to transmitter. This sensitivity increases with time
after initial learning and then declines. The rate at which such
sensitivity increases depends on the amount of initial learning.
If the curve of transmission plotted against time is displaced up-
wards with anticholinesterases then the very low portions will show
facilitation and the high portions will cause block (Figure 8). The
middle portions will appear unaffected (unless special experimental
tests are made). If the curve of transmission is displaced down
with anticholinergics then the middle portion will appear unaffected
and only the very early or late components will show block.

Taken together, then, the results which have been obtained are
evidence that synaptic conductance alters as a result of learning.
So far it seems: (1) that cholinergic synapses are modified as a
result of learning and that it probably is the postsynaptic membrane
which becomes increasingly more sensitive to acetylcholine with
time after learning up to a certain point; (2) After this point,
sensitivity declines, leading to the phenomena of forgetting; (3)
There is also good evidence that there is an initial phase of declining sensitivity to cholinesterase or increasing sensitivity to
anticholinergics. This could reflect the existence of a parallel
set of synapses with fast decay serving as a short term store; (4)
Increasing the amount of learning leads to an increase in conductance in each of a set of synapses without an increase in their
number; (5) Both original learning and extinction are subserved by
cholinergic synapses.

REFERENCES

Anderson, A.C. 1940. Evidences of reminiscence in the rat in
  maze learning. J. Comp. Physiol. Psychol., 30, 399-412.
Bacq, Z.M. and Brown, G.C. 1937. Pharmacological experiments on
  mammalian voluntary muscle in relation to the theory of chemical
  transmission. J. Physiol., 89, 45-60.
Beswick, F.B. and Conroy, R.T.W.L. 1965. Optimal tetanic conditioning of heteronymous monosynaptic reflexes. J. Physiol., 180.
  134.
Biederman, G.B. 1970. Forgetting of an operant response: Physostigmine produced increases in escape latency in rats as a
  function of time of injection. Quart. J. Exp. Psychol., 22, 384-
  388.
Bunch, M.E. and Lang, E.S. 1939. The amount of transfer of train-
  ing from partial learning after varying intervals of time. J.
  Comp. Physiol. Psychol., 27, 449-459.
Bunch, M.E. and Magdsick, W.K. 1933. The retention in rats of an
  incompletely learned maze solution for short intervals of time.
  J. Comp. Physiol., 16, 385-409.
Carlton, P.L. 1969. In: Reinforcement and Behavior, J.T. Tapp,
  Ed. Academic Press, New York.
Cutting, W.C. 1964. Handbook of Pharmacology: The Actions and
  Uses of Drugs, Appleton-Century-Crofts. New York.
DeGroot, J. 1957. Verhandel. Konink. Ned. Akad. Wettenschap.
  Afdel-Natuurk. Sect. II, 52.
Deutsch, J.A., Hamburg, M.D. and Dahl, H. 1966. Anticholines-
  terase-induced amnesia and its temporal aspects. Science, 151,
  221-223.
Deutsch, J.A. and Leibowitz, S.F. 1966. Amnesia or reversal of
  forgetting by anticholinesterase, depending simply on time of
  injection. Science, 153, 1017.
Deutsch, J.A. and Leibowitz, S.F. 1967. Memory enhancement by
  anticholinesterase as a function of initial learning. Nature,
  213, 742.
Deutsch, J.A. and Rocklin, K. 1967. Amnesia induced by scopolamine
  and its temporal variations. Nature, 216, 89-90.
Deutsch, J.A. and Stone, J. Unpublished.
Deutsch, J.A. and Wiener, N.I. 1969. Analysis of extinction
  through amnesia. J. Comp. Physiol. Psychol., 69, 179-184.
Eccles, J.C. 1961. The effects of use and disuse on synaptic
  function. In: Brain Mechanisms and Learning, ed. J.F. Dela-
  fresnaye, 335-352, Oxford, Blackwell Sci. Pub.
Eccles, J.C. 1964. The Physiology of Synapses. Berlin-Gottingen-
  Heidelberg: Springer.
Feldberg, W. and Vartiainen, A. 1934. Further observations on
  the physiology and pharmacology of a sympathetic ganglion. J.
  Physiol., 83, 103-128.
Fentress, J. and Doty. R. 1965. Protracted tetanization of the

optic tract in squirrel monkeys. Fed. Proc., 25, 573.

Goodman, L.S. and Gilman, A. 1965. The Pharmacological Basis of Therapeutics, New York, MacMillan.

Hamburg, M.D. 1967. Retrograde amnesia produced by intraperitoneal injection of physostigmine. Science, 156, 973-974.

Hubbert, H.B. 1915. The effect of age on habit formation in the albino rat. Behav. Monog. 2, no. 6.

Huppert, F.A. personal communication.

Huppert, F.A. and Deutsch, J.A. 1969. Improvement in memory with time. Quart. J. Exp. Psychol., 21, 267-271.

Liebowitz, S.F., Deutsch, J.A. and Coons, E.E. In preparation.

Rocklin, K. and Deutsch, J.A. Unpublished.

Russell, W.R. and Nathan, P.W. 1946. Traumatic amnesia. Brain, 69, 280-300.

Sharpless, S.K. 1964. Reorganization of function in the nervous system---use and disuse. Annual Rev. Physiol., 26, 357-388.

Spencer, W.A. and Wigdor, R. 1965. Ultra-late PTP of monosynaptic reflex responses in cat. Physiologist, 8, 278.

Squire, L.R. 1970. Physostigmine: Effects on retention at different times after brief training. Psychon. Sci., 19, 49-50.

Talland, G.A. 1965. Deranged Memory. Academic Press, New York.

Tanzi, E. 1893. Sulla presenza di cellule gangliari nelle radici spinali anteriori del gatto. Riv. sper. di Freniatria, 19, 373-377.

Volle, R.L. and Koelle, G.B. 1961. The physiological role of acetylcholinesterase (AChE) in sympathetic ganglia. J. Pharm. Exper. Ther., 133, 223-240.

Wiener, N.I. and Deutsch, J.A. 1968. The temporal aspects of anticholinergic and anticholinesterase induced amnesia for an appetitive habit. J. Comp. Physiol. Psychol., 66, 613-617.

SLOW BIOLOGICAL PROCESSES IN MEMORY STORAGE AND "RECOVERY" OF MEMORY

S. H. Barondes and L. R. Squire

Department of Psychiatry, University of California at

San Diego, La Jolla, California 92037

Despite all the evidence which has been reviewed at this symposium, the hypothesis that RNA and protein synthesis are required for long-term memory storage has not yet been firmly established. Although it is difficult to conceive of long-term memory storage in the brain without mediation of the prime cellular regulatory mechanisms which rely heavily on the synthesis of RNA and proteins, the various experiments which have been described here do not prove conclusively that these macromolecules are required. One could argue, for example, that increased incorporation of radioactive precursors into brain macromolecules during a learning experience is attributable to generalized cerebral activation or to changes in the specific radioactivity of precursor pools. Likewise, studies with inhibitors of cerebral protein synthesis can be faulted because of the known side effects of both puromycin (Cohen, Ervin and Barondes, 1966) and cycloheximide (Segal, Squire and Barondes, 1971). However, though we may challenge the interpretation of the experiments which have been presented, the general hypothesis remains supported.

If this hypothesis is so plausible, and if techniques are available for testing it, why the ambiguity? Several aspects of the memory problem complicate the investigation of its biological basis. The problem of behavioral classification, for example, arises when one attempts to describe any brain-behavior relationship. Thus, control groups designed to assess the role of non-learning factors in radioactive labelling experiments are only so adequate as our understanding of what factors constitute the learning process itself. In addition, available techniques are indirect and gross, considering the kind of subtle alterations in cellular connectivity which might

207

underlie information storage. Another aspect of memory which has
made consideration of its neural substrate perplexing is that the
memory storage process, unlike many biological processes which have
been studied, appears to develop slowly over a period of days.
This aspect of memory is the basis for the present discussion.

A SLOW PROCESS IN MEMORY STORAGE

Because memory storage is measurable immediately after learning, it is easy to conceive of it being stored by a rapid process
which is completed almost immediately and which lasts for a long
time. Yet a number of experiments indicate that slowly developing
biological processes may participate in memory storage. This
circumstance is not unique; biological systems do sometimes respond
to perturbations with complex, slow processes. In the extreme case,
combination of sperm cell and ovum triggers a series of events which
develop over months, years, and even decades. A less extreme example is the development of antibodies to a new antigen, which requires
complex cellular cooperation and cell division and which becomes
apparent several days after the antigen is introduced.

There are two lines of evidence to indicate that a slow process
is similarly involved in long-term memory storage. The first comes
from the changing responsiveness of acquired memory to treatments--
cholinesterase inhibitors (Deutsch, this volume), puromycin (Flexner,
Flexner and Stellar, 1963), actinomycin-D (Squire and Barondes,
1970), hippocampal surgery (Uretsky and McCleary, 1969)--during a
period of a week or more after training. The second comes from the
recovery of memory over a period of days to a week observed in some
studies following administration of glutarimide antibiotics.

Recovery of memory was first reported by Flexner et al. (1966)
in their initial report of the effects of acetoxycycloheximide on
memory. They found that if acetoxycycloheximide was injected intracerebrally immediately after training, memory was impaired for at
least 36 hours and then recovered. For several reasons, there was
a tendency to discount this recovery of memory. First, it was
clear that the animals were sick for at least one day after intracerebral injection of this long-acting antibiotic. Therefore, it
seemed possible that memory could have been present, but that sickness interfered with performance. When the sickness abated, performance improved. Second, our own work (Barondes, 1970) and that
of Agranoff (this volume) indicated that permanent amnesia could
indeed be produced with glutarimide antibiotics. In our experiments
permanent amnesia was produced if cycloheximide or acetoxycycloheximide was given shortly before training. The amnesia was in no
way related to performance, since 1) amnesia was observed long
after the animals recovered from the toxicity of the drug and 2)
administration of the drug to mice 30 minutes after training had

absolutely no effect on memory, measured either when the animals appeared sick or at any time thereafter.

In the past year, the recovery phenomenon has received renewed attention. Quartermain and McEwen (1970) reported that, when mice were give cycloheximide before passive avoidance training, conducted with a relatively high intensity footshock (1.6 mA), amnesia was observed 1 day but not 2 days after training. Although they did not report whether or not injections of cycloheximide after training affected memory one day later (thereby controlling for effects of the drug on performance), our own experience would indicate that side-effects of cycloheximide would not be expected to occur for more than a few hours after subcutaneous injection. Serota has also reported recovery of memory following acetoxycycloheximide (1971). He found impaired memory at one day, but recovery at seven days. Finally, Daniels (1971) has reported that acetoxycycloheximide-treated rats exhibit improvement in memory seven days after training. In our own experiments with a new automated discrimination task, the Deutsch Carousel, we have found that mice injected subcutaneously with cycloheximide 30 minutes before training are markedly amnesic 1 day after training, but show gradual recovery of memory over the following days (Squire and Barondes, in press) (Figure 1). The amnesia at 1 day is not a performance artifact, since injection of cycloheximide 30 minutes after training did not impair memory one day later. Our findings of recovery at seven days in the Deutsch Carousel contrast strikingly with our own experience with the T-maze (Barondes and Cohen, 1967) and with our studies of passive avoidance conducted with low shock (.35 mA) in which no measurable recovery occurred (Geller, Robustelli, Barondes, Cohen and Jarvik, 1969).

The critical aspect of this recovery is that there is a long period of amnesia (several days) before it occurs. Memory lasts for several hours, then is not demonstrable for days, then gradually reappears. The fact that the amnesia is transient has led some to argue that this deficit is due to a defect in "retrieval" rather than to impaired memory storage (Quartermain and McEwen, 1970). Yet, the drug impairs memory only if the inhibitor is present and active during training. Thus, the results cannot be due to a general disruption of retrieval or to a non-specific impairment of performance because a non-specific effect of the drug should be detectable one day after injection, whether the drug is given before training or shortly after training. The fact that the deficit does indeed depend on when the inhibitor is given relative to training strongly suggests that temporary amnesia is due to interference with some aspect of the memory storage process which occurs during training and/or within 30 minutes thereafter. If the storage process is impaired by protein synthesis inhibition, how does memory recover?

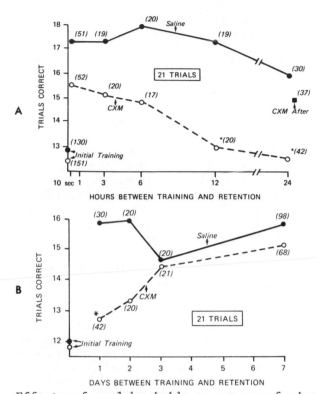

Figure 1. Effects of cycloheximide on memory of mice tested for
retention at different times after training. Mice were give 21
training trials in the Deutsch Carousel (Squire and Barondes, in
press). The initial training scores for mice given cycloheximide
or saline were nearly identical. Retention was tested at various
times later by giving an additional 21 trials. Numbers in parentheses indicate the number of mice in each group. The group
tested for retention at 24 hours appears in both figures. A) mice
were given cycloheximide or saline 30 minutes before training and
tested 10 sec, 1 hour, 3 hours, 6 hours, 12 hours, or 24 hours
later. An additional group was given cycloheximide 30 min after
training (CXM-AFTER) and tested 25 hours after training. The 24
hour group given CXM is significantly different from the CXM-AFTER
group (p < .01, t = 3.0). Every group given CXM is significantly
different from the corresponding group given saline (p < .05).
B) mice were given cycloheximide or saline 30 minutes before training and tested 1 day, 2 days or 7 days later. Mice given saline
performed better than mice given CXM at 1 day (p < .001, t = 4.1)
and at 2 days (p < .05, t = 2.4) after training. The 7 day group
tested with CXM exhibited significant savings (p < .001, t = 4.7)
and was significantly different from mice treated with CXM and tested
1 day (p < .001, t = 4.2) or 2 days (p < .01, t = 2.5) after training. *Not significantly different from the original training score
for this group.

RECOVERY OF MEMORY: THREE HYPOTHESES

1. The Stable Messenger-RNA Hypothesis

In considering their finding of recovery, Flexner et al (1966) proposed the ingenious explanation that it was mediated by stable messenger RNA. In this view, cerebral protein synthesis is required for expression of memory. With learning, messenger RNA is synthesized which directs this protein synthesis. Normally, this stable messenger is translated, producing the protein required for expression of memory. Long-term memory is thereby readily established. In the presence of acetoxycycloheximide, a long-acting inhibitor of cerebral protein synthesis, it was postulated that stable messenger-RNA synthesis continues normally, but its translation is blocked by the inhibitor. When the inhibitor disappears the stable messenger-RNA is translated and memory gradually appears. This hypothesis is consistent with the data of Flexner's original experiments. It seemed particularly appealing because the glutarimide antibiotics are known not to promote the destruction of previously synthesized messenger-RNA. The major difficulty with this hypothesis is the lag between cessation of cerebral protein synthesis inhibition and recovery of memory which is observed in our experiments (Figure 1). Protein synthesis inhibition by cycloheximide has terminated within about 4 hours of administration of the drug. At that time, memory is still exhibited. Amnesia does not develop until 12 hours after training, and it persists for at least 2 days. Were a stable messenger-RNA made during training, one would expect that its translation would begin as soon as the action of cycloheximide has ceased. Why then does memory not recover for several days?

2. Slow Axoplasmic Transport

In seeking an explanation for slow development of memory in cycloheximide-treated animals, it is natural to seek a process which normally develops slowly in the brain. The only such process which has clearly been shown to occur in the mouse brain is slow axoplasmic transport. It is now clearly established that proteins synthesized in nerve cell bodies are transported to nerve terminals at two major rates--a slow rate in the range of several mm/day and a rapid rate in the range of 400 mm/day (McEwen and Grafstein, 1968; Ochs, Sabri and Ranish, 1969). In the mouse brain, it has been clearly shown that labelled proteins are detectable at nerve terminals within 15 minutes after injection of radioactive amino acid (Droz and Barondes, 1969). However, some classes of proteins, notably certain soluble proteins, arrive at the nerve endings over a period of days (Barondes, 1968; Table I). How might slow axoplasmic transport be related to the recovery process which we are considering? Perhaps rapidly transported proteins or proteins synthesized by the ribosome-containing postsynaptic dendrites are involved in memory for the first few days after training; and

TABLE I

Specific Activity of Soluble Protein from Whole Brain
and from Nerve Endings at Different Intervals of Time
after Injection

Counts/Min/Mg Protein

Time after Injection	Soluble Protein of Whole Brain	Soluble Protein of Nerve Endings	Ratio*
1 hr	2940	188	0.064
2 hrs	3020	248	0.082
4 hrs	2980	322	0.108
6 hrs	2870	370	0.129
16 hrs	2650	562	0.212
24 hrs	2490	667	0.268
4 days	1320	715	0.542
8 days	770	793	1.030
16 days	390	554	1.420
23 days	280	423	1.510

Mice were injected intracerebrally with ^{14}C-leucine and sacrificed
at the indicated times after injection. Subcellular fractions were
prepared from homogenates of two cerebral hemispheres. The soluble
fraction of whole brain is that portion which does not sediment
after centrifugation at 100,000 x g for 1 hour. The soluble fraction of nerve endings is that fraction of purified nerve endings
which, after lysis of the particles with water, does not sediment
at 100,000 x g for 1 hour.

*Ratio of soluble protein of nerve endings to soluble protein of
whole brain.

proteins slowly transported to the presynaptic region are required
to strengthen synaptic connnections over periods which are longer
than that. The outstanding weakness of this notion is that the
synthesis of proteins transported to nerve endings by slow axoplasmic transport, such as microtubular protein (Feit, Dutton, Barondes
and Shelanski, 1971), is known to be inhibited as extensively by
cycloheximide as those transported to nerve endings by rapid axoplasmic transport. The idea remains tenable, however, if one
assumes that the slowly transported protein is synthesized far more
redundantly than rapidly transported protein. In this view, the
slowly transported protein, although its synthesis is strikingly
inhibited by cycloheximide, is made in such enormous excess that
enough is spared to mediate memory when it arrives at appropriate

nerve terminals some days later. Although this reasoning is not
very parsimonious, the idea does have the potential merit of being
testable. There are indeed drugs like colchicine and vinblastine
which inhibit axoplasmic transport (Sampson, 1971). Should it be
possible to inhibit axoplasmic transport without threatening survival or grossly disrupting the behavior of the animals (a goal which
we have not yet been able to achieve), the hypothesis could be
tested. Its usefulness remains to be determined.

3. An "Autocatalytic" Process in Memory Storage

Perhaps because of its vagueness, the third alternative seems
presently most appealing. It attempts to relate normal "maintenance"
of memory with the slow recovery of memory observed in cycloheximide-
treated mice. Maintenance is a major problem in the study of memory.
Not only does the memory process develop over some period of time,
but it persists for years. Since the molecules which mediate memory would be expected to turn over with half lives in the range of
days or weeks, how is memory maintained? The vague position to
which one is forced is that the change which mediates memory storage
somehow induces a cellular condition which leads to continued biosynthesis of this change (Barondes, 1965; Roberts and Flexner, 1969).
This process probably develops over days, since the half-lives of
the macromolecules involved would be expected to be at least that
long. Since the glutarimide antibiotics do not totally inhibit
cerebral protein synthesis, a small amount of the protein required
for memory storage might be synthesized even in animals whose cerebral protein synthesizing capacity is markedly inhibited. In one
limiting case, with overtraining (Barondes, 1970; Squire and Barondes,
in press), there is apparently enough synthesis for full development
of memory without any period of amnesia. In the other limiting case,
with undertraining (Barondes, 1970), there may be too little synthesis to lead to memory storage even when measured long after training.
In the intermediate case, possibly represented by the studies in
Figure 1, a small amount of synthesized protein, although not sufficient for expression of memory one day after training, may be sufficient to initiate the autocatalytic process which is postulated to
be normally required for the maintenance of memory. As the autocatalytic process augments, memory gradually appears.

According to this view, the gradual appearance of memory during
the days after training may be related to the process which normally
makes memory gradually invulnerable to a variety of agents, such as
puromycin (Flexner, Flexner and Stellar, 1963) or actinomycin-D
(Squire and Barondes, 1970) during the days after information has
been acquired. In the normal case, proteins synthesized at the time
of training induce an autocatalytic process of sufficient initial
strength so that by 7 days after training this process is relatively
unperturbable. In the case of training during protein synthesis

inhibition, however, only a small amount of synthesized protein is
initially available to initiate this process. As a result, it
develops slowly, and several days are required before memory can be
expressed.

This hypothesis might be tested by comparing the effects of
amnesic treatment at 7 days after training in normal animals and
in animals trained during protein synthesis inhibition. If cycloheximide-treated mice are characterized by a slowly developing memory storage process, these animals might at 7 days after training
still be susceptible to those amnesic treatments which can affect
normal animals only when given within 1 day after training. According to this idea, recovery of memory observed at 7 days in cycloheximide-treated mice may not represent complete reversal of the
amnesic effect. Instead, it may reflect memory storage that is sufficient for expression, but which (as in normal mice 1 day after
training) requires still further development to become resistant to
perturbation. The hypothesis being considered here might also be
tested by attempting to produce prolonged inhibition of cerebral
protein synthesis during the period when the presumed autocatalytic
process is occurring. Since this procedure will produce severe
illness, this strategy promises to have only limited success.

SUMMARY

A major weakness of the hypothesis that cerebral protein synthesis is required for long term memory storage comes from the finding that in some experimental situations mice trained while 95% of
cerebral protein synthesis is inhibited by glutarimide antibiotics,
have impaired memory for only a few days. Thereafter, memory progressively recovers. It has been pointed out that both the development of memory storage and also the maintenance of memory appear
to involve slow biological processes. Possible mechanisms of
recovery are considered which involve stable messenger RNA, slow
axoplasmic transport, or an autocatalytic process requiring protein
synthesis.

ACKNOWLEDGMENT

Supported by Grant MH-18282 from the U.S.P.H.S.

REFERENCES

Agranoff, B.W. Further studies on memory formation in the goldfish.
  This volume.
Barondes, S.H. 1965. Relationship of biological regulatory mech-
  anisms to learning and memory. Nature, 205, 18-21.
Barondes, S.H. 1968. Further studies of the transport of protein
  to nerve endings. J. Neurochem., 15, 343-350.

Barondes, S.H.   1970.   Cerebral protein synthesis inhibitors block
    long-term memory. Internat. Rev. Neurobiol., 12, 177–205.
Barondes, S.H. and Cohen, H.D.   1967.   Delayed and sustained effect
    of acetoxycycloheximide on memory in mice. Proc. Nat. Acad. Sci.
    14, 371–376.
Cohen, H., Ervin, F. and Barondes, S.H.   1966.   Puromycin and cyclo-
    heximide: Different effects on hippocampal electrical activity.
    Science, 154, 1557–1558.
Daniels, D.   1971.   Acquisition, storage, and recall of memory for
    brightness discrimination by rats following intracerebral infusion
    of acetoxycycloheximide. J. Comp. Physiol. Psychol., 76, 110–118.
Deutsch, J.A.   The cholinergic synapse and the site of memory. This
    volume.
Droz, B. and Barondes, S.H.   1969.   Nerve endings: Rapid appearance
    of labeled protein shown by electron microscope radioautography.
    Science, 165, 1131–1132.
Feit, H., Dutton, G.R., Barondes, S.H. and Shelanski, M.L.   1971.
    Microtubule protein: Identification in and transport to nerve
    endings. J. Cell Biol., 51, 138–147.
Flexner, L.B., Flexner, J.B. and Roberts, R.B.   1966.   Stages of
    memory in mice treated with acetoxycycloheximide before or
    immediately after learning. Proc. Nat. Acad. Sci., 56, 730–735.
Flexner, J.B., Flexner, L.B. and Stellar, E.   1963.   Memory in mice
    as affected by intracerebral puromycin. Science, 141, 57–59.
Geller, A., Robustelli, F., Barondes, S.H., Cohen, H.D. and Jarvik,
    M.E.   1969.   Impaired performance by post-trial injections of
    cycloheximide in a passive-avoidance task. Psychopharmacologia,
    14, 371–376.
McEwen, B.S. and Grafstein, B.   1968.   Fast and slow components in
    axonal transport of protein. J. Cell Biol., 37, 494–508.
Ochs, S., Sabir, M.E. and Ranish, N.   1969.   Somal site of synthe-
    sis of fast transported materials in mammalian nerve fibers. J.
    Neurobiol., 1, 329–344.
Quartermain, D. and McEwen, B.S.   1970.   Temporal characteristics
    of amnesia induced by protein synthesis inhibitor: Determination
    by shock level. Nature, 228, 677–678.
Roberts, R.B. and Flexner, L.B.   1969.   The biochemical basis of
    long-term memory. Quart. Rev. Biophys., 2, 135–173.
Sampson, F.   1971.   Mechanism of axoplasmic transport. J. Neurobiol.
    2, 347–360.
Segal, D.S., Squire, L.R. and Barondes, S.H.   1971.   Cycloheximide:
    Its effects on activity are dissociable from its effect on memory.
    Science, 172, 82–84.
Serota, R.G.   1971.   Acetoxycycloheximide and transient amnesia in
    the rat. Proc. Nat. Acad. Sci., 56, 730–735.
Squire, L.R. and Barondes, S.H.   1970.   Actinomycin-D: Effects on
    memory at different times after training. Nature, 225, 649–650.
Squire, L.R. and Barondes, S.H.   Variable decay of memory and its re-
    covery in cycloheximide-treated mice. Proc. Nat. Acad. Sci., in

press.

Uretsky, E. and McCleary, J. 1969. Effect of hippocampal isolation
  on retention. J. Comp. Physiol. Psychol., 68, 1-8.

DISCUSSION

Professor David Krech

Department of Psychology, University of California,

Berkeley, California

I have been accused of many, many things, Professor McGaugh, but never of being an Unpolarizer! Au contraire . . .

You may remember that Dr. Agranoff, early this morning--several hundred mice, goldfish and macromolecules ago--started with a note on the weather. He said that every morning when he got to this Conference Hall he found the atmosphere rather foggy, and that this morning (with his own paper on the program) it promised to be the most foggy day of all. I must admit that Dr. Agranoff is a very good weather prophet, but had he spent a few more days in the Bay Area, he would have learned that inevitably later in the day a hot, bright, sun comes through and burns off the fog, making everything crystal clear. And I assume, Dr. McGaugh, that that is my function now. You must therefore expect a few searing re-marks, but the heat in these remarks is intended to shed light.

I will start with a few comments directed specifically at the papers we heard this morning, and then I may get around to some general points.

I start with Glassman. His study was a good example of cer-tain kinds of research which seek to discover changes in brain chemistry induced by training or experience. I must confess that I find it difficult to understand the experimental rationale or to take seriously the results of his kind of research. Here is why. Typically, in such studies, the experimental animal is given some specified training, preferably (if the experimenter is a good and dedicated biochemist or physiologist) training which requires a minimum of experimental skill or time on the part of the laboratory

217

personnel. Ideally, the experimenter seeks a training procedure
which, say, takes no more than five minutes a day and can be carried out by the least-trained technician who happens to be available. This, you see, will permit the expenditure of skill and time
where it really counts, on the chemistry or neurophysiology. And,
if you are a good experimenter (and Glassman is an excellent experimenter) you also have the necessary control animals. "Control
animals" are defined as animals which are not given this minimum of
training, or, if you are really sophisticated (and Glassman is
really sophisticated) you have a "yoked control," that is, an
animal which gets the same shocks which the experimental animal
receives (always, it seems, electric shocks must be involved in a
scientific experiment having to do with memory!) but cannot learn
how to avoid these shocks. Finally, you sacrifice your animals
and look for differences in their brain chemistries.

Now here is the root of my difficulty. If what we are looking
for are the effects in brain chemistry which somehow have something
to do with the living animal's experience, how in God's name can we
possibly expect to detect such effects from so miniscule and con-
fusing a difference in the experiences of our experimental and
control animals. Look here, these animals (at least in our lab)
live all 24 hours a day. During these 24 hours these animals (controls as well as experimentals) are solving problems (rat-type
problems, to be sure), or sleeping, or fighting, or lusting, or
scratching. They are behaving, they are experiencing, and presumably all kinds of chemical residuals are being stored up in their
brains--24 hours per day--as a consequence of these behaviors and
experiences. How can Glassman expect a 5-minute differential
experience to lay down differentiated chemical residuals in active,
living brains; residuals which could then be detected and differentiated by necessarily crude chemical post-mortem scouting expedi-
tions in those brains? This expectation becomes more wondrous to
me when one uses (as Glassman did) a yoked control. Specifically:
Did not Glassman's yoked animal experience the shock? Was not the
yoked animal reminded of the shock when it was placed in the "reminder" cage? Or will Glassman require that we suspend logic and
believe that the yoked rat had forgotten its experience simply
because the "reminder" cage was intended to remind only the experimental animal? Buried here, it seems to me, is the basic fallacy
of these experimental designs. It is the fallacy which I shall
call "Experimentomorphism"--i.e., attributing to the perception and
memory of an animal only that which would pleasure the experimenter
to have the animal experience. More specifically, for Glassman's
experiment, we must assume that the animal will learn and will
remember, and that its brain chemistry will undergo appropriate
chemical changes to, only those events which the experimenter, by
virtue of the authority somehow vested in him, has designated as
relevant. All other experiences are to be shunned by the animal's

psyche and brain as null and void and without shape, form or chemical effect. This is chutzpah indeed! And it is because I find it
difficult to make the necessary experimentomorphic assumptions--for
such assumptions go against all my logic and all my wisdom--that I
find it difficult to take seriously the resulting data of such
studies. Of course, if such studies should yield consistent and
replicable data (and the history of science is replete with instances
where experiments have yielded compelling data against all logic
and wisdom) then I promise I will relent my stiff-necked opposition
and yield gracefully to a cumulative record. But thus far I have
seen no need to be graceful.

I turn now to Agranoff who takes a different approach to our
problem. He wants to determine the effects of various chemical
agents upon memory. Agranoff knows how to manipulate his chemical
agents, but how about the variable of memory? And it is here that
I begin to feel uneasy. You remember perhaps that Agranoff started
out by flattering us with the assurance that there was no need for
him to spell out the reasons for choosing the goldfish as his
experimental animal; it has, he pointed out, the many obvious ad-
vantages for biochemical and surgical-intervention studies with
which we were all familiar. But far from being flattered, I was
dismayed! Look here, Agranoff is working, as we all are, with two
quite different (at least phenomenally) families of variables; the
physiological or biochemical variables, and the behavioral or
psychological ones. And anyone who is to do a good study involving
these two sets of variables had damn well better be sophisticated,
and careful, and disciplined, and knowledgeable and thorough in
both sets of variables. If this be so, one does not pick a goldfish because the goldfish is good for manipulating and assessing
one set of variables--and then disregard the animal's appropriate-
ness for the other set. To do this is to run the risk of ending
with a half-assessed experiment. We must seek an animal which is
the animal of choice for both families of variables. There is no
way out of it. I know it is a tough requirement, but there is no
way out of it.

I am reminded of that old saw about the New Englander who,
when asked the way to St. Johnsbury by a tourist said, "Well, you
might go down that left fork and then . . . no, you'd better not,
that road's been washed out. I'll tell you, go left here and . . .
no, the old bridge won't hold your car. You know, if I were you,
I wouldn't start from here!" Now, that is not very helpful advice
to give to a man who is here and wants to get to St. Johnsbury.
But it may be excellent advice for Agranoff. He does not have to
start with the goldfish; if the goldfish is not good for studies
of memory, or behavior, or intelligence, he can start with some
other beast. But in any event it is incumbent upon him to justify
the use of the goldfish for memory studies.

One more comment about Agranoff's paper. Every student of psychology knows that to measure memory is an extremely subtle and difficult and at the same time deceptively easy thing to do. There are at least three traditional ways of measuring memory in man, ways which have been known for about 100 years now: the reproductive, the recognition, and the relearning methods. And each method may give quite different results! And this is true for both man and the lower animals--for there, too, have these measurement methods been tried. Now, which one of these methods measures "memory"?

It seems to me that anyone working on the brain chemistry of "memory" cannot possibly be content with one measure of memory if he wants to make some useful assertions about the effect of this or that chemical agent on the laying down of memory traces. Agranoff is too good a scientist to be cavalier in his handling of so important a variable of his research design.

Now to Deutsch. Deutsch's study does not quite fit my psychologically-biased harangue on this afternoon of my discontent. Where Agranoff says, "I will manipulate chemical agents in order to see the effects of these agents on memory," Deutsch says, "I will manipulate memory and study the effects of my chemicals on the memory which I manipulate." In doing so, Deutsch, of course, is displaying his psychological training and orientation. And so we find Deutsch studying the natural history of a memory--what happens to a memory at different stages of its development. He even went way, way back to the old psychological concept of reminiscence in this effort (although Deutsch, for reasons not made clear, did not use the term nor show the continuity between his studies and these older, classical studies). I must say that I liked much of Deutsch's handling and analysis of his behavioral data--would that I could be as enthusiastic about his pharmacological and neurophysiological arguments.

Enough now, of these specific comments on this morning's papers. Let me end with a few general comments about the field wherein we all labor and about our labors therein. I am discouraged by what has happened to the field of brain chemistry and behavior. I have been in this field a long time and I seem no longer to sense the kind of shiny, bright optimism which was exuded by almost every study reported at the first AAAS symposium in Berkeley where many of us were present so long, long ago. It seems to me that the field is getting tired--or timid--or tired _and_ timid _and_ encapsulated. I see the same faces, the same data, and the same errors again and again and again. The field is now about 20 years old; old enough to have a history, and the history is taking on the outlines of the awful histories of some of the other areas of psychology with which I am familiar; for instance, the history of verbal learning (up to this decade) or the history of learning theory in general. What

happened in the latter instance was that in the beginning, learning
theorists sought to write a cosmic theory of learning--this was,
say, during the twenties and thirties of this century. They were
going to cover everything, from how a rat learns to discriminate
between a black and white card to the creative thinking of an
Einstein. But they soon saw that this was a big order--especially
if the theory was to be supported by empirical studies. Heavens!
One had to know (i.e., discover) a great deal about how people
thought, how rats learned, and how cockroaches forgot! What to do?
They might have said, "This is a tough order, let us try it." Well,
a very few did, but most real "scientific" theorists said, "Let us
cut our problem down to size." Experimental apparatus was simpli-
fied--instead of complex mazes for rats, why not use single-unit
running alleys, for example; the variety of species examined was
sharply restricted--why not limit ourselves to the good old laboratory rat; behavior observations were restricted--observe only the
minimum necessary for "objective" records. And soon it came to pass
that almost all of the learning theory was based on studies of the
rat, or preferably the pigeon, encased in a sound-proof, light-
proof, experimentor-proof, concept-proof box where the pointer-
readings of the encapsulated pigeon's pecking or the rat's jabbing
at a lever were the only behaviors (?) recorded or considered.
Not only were the animals encapsulated, so were the theory and the
thinking of the theorist. Almost never did the theorist look at or
worry about how a lively, defecating, urinating, copulating, biting,
rat learns his way about in nature, or how an Einstein thinks. That
is, never except when the theorist extrapolated from his pigeon
(which he had never seen but which he had "studied" intensively) to
man (whom he may have seen but had never studied).

     In our present field, we all agree that we are interested in
memory. Memory is a big and rich and awesome phenomenon. And
psychologists already know quite a bit about memory--but much of
this knowledge lies scorned and unused because we have cut down
our field to size. Barondes has just told us that introspection
tells us that memory is laid down immediately, and he is therefore
puzzled because his (and others') brain chemistry studies seem to
suggest that the laying down of memory is prolonged over time. But,
Dr. Barondes, psychological studies of the last 40 to 50 years tell
us that the laying down of memory is a dynamic, long-term process.
Not only is it a long-term affair, but it is a many-splendoured,
qualitatively-changing thing. And this the psychologists who have
studied memory in man have long known--ask Bartlett, as Kohler, ask
Wulf, ask Wertheimer. Barondes and Glassman and Agranoff and
Deutsch and all of the rest of us are never going to find out much
about memory and brain chemistry unless we pay a decent respect to
what is already known about memory in man. We dare not encapsulate
ourselves in the goldfish, or mouse, or rat, or even the human
laboratory. We must look at people--raw, living unyoked people.

Dr. McGaugh, anticipating this criticism, told us in his introductory remarks this afternoon that our research is not encapsulated. He points out that our interest and experiments in consolidation theory, as one example, derive from clinical observations of retrograde amnesia in man. True enough, it may have been that those observations sent us into the laboratory, but what we haven't done--at least sufficiently--is gotten <u>out</u> of the laboratory again. We have to observe people, and then not only go into the laboratory to check out our observations (and there we may use goldfish perhaps, or mice, or rats), but we also have to get out of the lab and check out our experiments on people.

Here I want to propose the formation of a society which may save us yet, and I invite all of you to become charter members thereof. I shall call this the Society of St. Anteus. You do not know who St. Anteus is because there is no such Saint. But there is an Anteus in Greek mythology. It seems that Anteus was marvelously strong and loved to wrestle. His challenge to the world was straightforward: "Let us wrestle unto the death--no compromises!" And Anteus always won until Hercules came along. The story goes that the mighty Hercules picked Anteus up, twirled him around (à la the television wrestlers) and slammed Anteus down on the ground. And lo! Anteus jumped up from the ground with renewed vigor and continued the match. For, you will remember, Anteus was the son of Mother Earth, and every time he touched Earth, he regained full vigor and renewed zest. It was when Anteus remained out of touch with the soil from which he had sprung that he began to wilt. Clever Hercules soon perceived this and held Anteus aloft--as far above the Earth as he could get him--and choked Anteus until Anteus was gone. And so was Anteus martyred and as far as I am concerned, sainted, since by his martyrdom Anteus doth teach us a most precious lesson.

Most of us here started our brain chemistry and behavior research with some people-based observation or intuition, or hunch. We then, being experimental scientists, went unto our rats and pigeons and mice and goldfish. We did the best we know how. Good. But, and this is my plea, let us venture forth again, let us be quicker than we have been to check out our rat-pigeon-goldfish-mouse findings on people. Let us not, in remaining away from man whence all of our first hunches flow, suffocate in our animal laboratories. Let us not repeat the dismal history of the learning theorists; let us not become like unto the baroque discipline of yesterday's verbal learner where we fret and worry to death the tiny little bits of detailed observations signifying nothing much about anything. Let us go back to man. And why don't we do that? I think it is because most of us suffer from a failure of nerve. We are afraid that it is too tough a task to manage, and we may even be afraid that without manageable pointer-readings we may lose our status as scientists.

Let me address a plea to my colleagues, the psychologists. You know what I think it means to be a good psychologist on a brain-behavior research team? It means knowing how to talk to people, how to listen to them, how to elicit from them good introspective, reports of memory. It means being a psychologist and clinician, and a teacher of behavioral science to the biochemist, and pharmacologist, and neurophysiologist, and anatomist. If you, psychologists, don't do this for your colleagues, who will? And if not now, when?

In one of the informal discussions during this afternoon's intermission the suggestion was made that the brain chemistry of mental disorders will advance more quickly than the brain chemstry of memory. There is merit in this suggestion if for no other reason than this: Many psychiatrists who work on brain chemistry and mental disorders have had intimate experience (even if vicariously) with mental disorders in working with their patients. They have been forced to listen and observe—carefully and with sympathy and even with empathy. On the other hand, many brain researchers who work with memory have not had occasion to take a serious, disciplined look at memory. I do not know what a memory is for the rat. I cannot empathize with my brother rat that much. But I think I do know something about what memory is in people. Well, then, why not *in* *addition* to working with the rat, why not also work with people?

I think that the next great breakthrough in this field, a breakthrough which must come if we are to see a revival of bright and promising work (and here I would like to quote Ghiselin: ". . . the crudest approximation, if it provides hints for the solution of a broad range of problems, has every advantage over the most elegant mathematical law which asserts nothing of interest.") will be a breakthrough according to the Gospel of St. Anteus. Break through your laboratory walls. Go constantly to look at memory in people. Know what you are studying, and whence came your question, and before whom your work will be judged. Take your laboratory findings and bring them out, and test them on people and return refreshed to wrestle with renewed zest with your pigeons and rats and mice. To do this will take scientific valor, experimental ingenuity, and a readiness to commit science, yea even among those who would do good, for it is conceivable that should you work with people, you might even find that you will do *well* with your rat work by doing *good* among the mentally retarded, the senile, the disturbed.

INDEX